'This new book by Timothy Foxon is a welcome addition to the growing literature on how to achieve a transition to a low-carbon economy. Dealing with basic research as well as policy questions, the book is unique in that it bridges energy, innovation and transition studies, while devoting also attention to the recently revived debate on the complicated relationship between economic growth and energy use. Eloquently written, ambitious in scope and synthetic in approach – it deserves a wide readership.'

– Jeroen van den Bergh, ICREA Research Professor at Universitat Autònoma de Barcelona, Spain

'In this thought-provoking, readable analysis, Timothy Foxon offers challenging new insights into pathways to more sustainable and prosperous energy and economic systems. Drawing on his deep knowledge of evolutionary economics, ecological economics and innovation, he illuminates the complex past and prospective future relationships between energy, economic growth and our environment.'

– Peter Pearson, Professor, Imperial College London, UK

'The links between energy and economic growth are well-known, yet Timothy Foxon manages to bring a fresh lens to the topic, particularly in the context of low carbon transformation and green growth. With highly accessible explanations, this book allows readers to more fully understand some of the larger debates of our times.'

– Kathleen Araújo, Assistant Professor, Stony Brook University, USA

ENERGY AND ECONOMIC GROWTH

Access to new sources of energy and their efficient conversion to provide useful work have been key drivers of economic growth since the industrial revolution. Western countries now need to transform their energy systems and move away from the single-minded pursuit of economic growth in order to reduce our carbon emissions, and to allow the environmental space for other countries to develop in a more sustainable way. Achieving this requires understanding of the dynamics of economic and industrial change with appreciation of the dependence of economies on ecological systems.

Energy and Economic Growth thus examines the links between three issues: history of energy sources, technologies and uses; ecological challenges associated with the current dominant economic growth paradigm; and the future low carbon energy transition to mitigate human-induced climate change. Providing a historical understanding of the relevant connections between physical, social and economic changes, the book enables the reader to better understand the connection between their own energy use and global economic and environmental systems, and to be able to ask the right questions of our political and business leaders.

This is a valuable resource for students, scholars and policy makers with an interest in energy, climate change and economic thinking.

Timothy J. Foxon is Professor of Sustainability Transitions at SPRU (Science Policy Research Unit), University of Sussex, UK. He previously held academic and research positions on sustainability and innovation at the University of Leeds, University of Cambridge and Imperial College London. He has a PhD in theoretical physics from the University of Cambridge.

ROUTLEDGE STUDIES IN ENERGY TRANSITIONS

Considerable interest exists today in energy transitions. Whether one looks at diverse efforts to decarbonize, or strategies to improve the access levels, security and innovation in energy systems, one finds that change in energy systems is a prime priority.

Routledge Studies in Energy Transitions aims to advance the thinking which underlies these efforts. The series connects distinct lines of inquiry from planning and policy, engineering and the natural sciences, history of technology, STS, and management. In doing so, it provides primary references that function like a set of international, technical meetings. Single and co-authored monographs are welcome, as well as edited volumes relating to themes, like resilience and system risk.

Series Editor: Dr. Kathleen Araújo, Stony Brook University, USA

Titles in this series include:

How Power Shapes Energy Transitions in Southeast Asia
A complex governance challenge
Jens Marquardt

Energy and Economic Growth
Why we need a new pathway to prosperity
Timothy J. Foxon

Accelerating Sustainable Energy Transition(s) in Developing Countries
The challenges of climate change and sustainable development
Laurence L. Delina

ENERGY AND ECONOMIC GROWTH

Why we need a new pathway to prosperity

Timothy J. Foxon

Routledge
Taylor & Francis Group

LONDON AND NEW YORK

earthscan
from Routledge

First published 2018
by Routledge
2 Park Square, Milton Park, Abingdon, Oxon OX14 4RN

and by Routledge
711 Third Avenue, New York, NY 10017

Routledge is an imprint of the Taylor & Francis Group, an informa business

British Library Cataloguing-in-Publication Data
A catalogue record for this book is available from the British Library

Library of Congress Cataloging-in-Publication Data
Names: Foxon, Tim, 1967- author.
Title: Energy and economic growth : why we need a new pathway to prosperity / Timothy Foxon.
Description: Abingdon, Oxon ; New York, NY : Routledge is an imprint of the Taylor & Francis Group, an Informa Business, [2018] | Series: Routledge studies in energy transitions
Identifiers: LCCN 2017024386 | ISBN 9781138669284 (hbk) | ISBN 9781138669307 (pbk) | ISBN 9781315618180 (ebk)
Subjects: LCSH: Economic development–Environmental aspects. | Sustainable development.
Classification: LCC HD75.6 .F69 2018 | DDC 338.9/27--dc23
LC record available at https://lccn.loc.gov/2017024386

ISBN: 978-1-138-66928-4 (hbk)
ISBN: 978-1-138-66930-7 (pbk)
ISBN: 978-1-315-61818-0 (ebk)

Typeset in Bembo
by Integra Software Services Pvt. Ltd.
Printed and bound by CPI Group (UK) Ltd, Croydon, CR0 4YY

We succeeded in taking that picture [from deep space] and if you look at it, you see a dot. That's here. That's home. That's us. On it, everyone that you ever heard of, every human being who ever lived, lived out their lives. The aggregate of all of joys and suffering, thousands of confident religions, ideologies and economic doctrines, every hunter and forager, every hero and coward, every creator and destroyer of civilizations, every king and peasant, every young couple in love, every hopeful child, every mother and father, every inventor and explorer, every teacher of morals, every corrupt politician, every superstar, every supreme leader, every saint and sinner in the history of our species, lived there in a mote of dust, suspended in a sunbeam.

Carl Sagan, 'Pale Blue Dot' speech, 1994

CONTENTS

ILLUSTRATIONS

Figures

Tables

PREFACE AND ACKNOWLEDGEMENTS

This book builds on my research over the last 20 years on understanding the potential for, and challenges of, a low carbon energy transition. Though the book is largely a solo effort, most of this research has been done in collaboration with colleagues and PhD researchers. So, this book would not have been possible without this collaboration (though, of course, I am responsible for all remaining errors in the text).

First of all, I am very grateful to my new colleagues at SPRU, the Science Policy Research Unit, at the University of Sussex, just outside Brighton, UK, for providing a congenial and intellectually stimulating new home. Thanks especially to Johan Schot, the Director of SPRU, and Steven McGuire, Head of the School of Business, Management and Economics, of which SPRU is part, for bringing me to Sussex and for providing support for the first two years of my position as Professor of Sustainability Transitions. This freed me from any heavy teaching or administrative responsibilities, and so directly enabled the writing of this book. I would like to thank colleagues at SPRU who provided direct feedback on drafts of a couple of chapters at a discussion session, including Johan Schot, Ariel Wirkierman, Andy Stirling, Phil Johnstone, Cian O'Donovan, Jenny Lieu, Laur Kanger and Chiara Frantini. I would also like to thank other colleagues at SPRU and the Institute for Development Studies at Sussex for interesting and enlightening conversations, including Rali Hiteva, Kat Lovell, Maria Savona, Steve Sorrell, Gregor Semeniuk, Benjamin Sovacool, Ed Steinmueller, Lucy Baker, Sabine Hielscher, Stephen Spratt, Pari Patel and Mariana Mazzucato. Thanks also to staff at SPRU and Sussex for easing my transition and providing a range of support, including Marion Clarke, Sarah Schepers, Charlotte Humma, Nora Blascsok, Katherine Davies, Jenny Bird and Pip Bolton.

Some of the ideas in this book were developed while I was previously at the University of Leeds. In particular, I am grateful for supportive collaborations with both Julia Steinberger and Lucie Middlemiss on working and conference papers that helped to elaborate these ideas. I would also like to thank other colleagues at my previous intellectual homes at the University of Leeds, University of Cambridge and Imperial College London for their support and encouragement, including Andy Gouldson, Jouni Paavola, Peter Taylor, Bill Gale, Lindsay Stringer, Anne Tallontire, Alice Owen, Susie Sallu, Paola Sakai and John Barrett at Leeds; Terry Barker, Şerban Scrieciu and Jonathan Köhler at Cambridge; and David Butler, Matt Leach, Rob Gross, Roger Fouquet and the late Dennis Anderson at Imperial College.

I am grateful for financial support from UK and EU taxpayers, through the UK Economic and Social Research Council (ESRC), the Engineering and Physical Sciences Research Council (EPSRC) and the European Commission Sixth Framework Programme. This includes current research funding as part of ESRC Centre for Climate Change Economics and Policy (CCCEP) (grant reference ES/K006576/1); EPSRC/ESRC i-BUILD: Infrastructure BUsiness models, valuation and Innovation for Local Delivery project (grant reference EP/K012398/1); UK Energy Research Centre Phase 3 (grant reference EP/L024756/1); and the Research Councils UK Centre for Innovation on Energy Demand (CIED) (grant reference EP/K011790/1). Other important leaders and collaborators on these and earlier research projects have included Simon Dietz, Judith Rees, Alex Bowen, Richard Dawson, Phil Purnell, Andy Brown, Jim Watson, Frank Geels, Tim Schwanen, Geoff Hammond, Neil Strachan, Paul Ekins, Jonathan Michie, Christine Oughton, Karsten Neuhoff, Liz Varga and Rene Kemp.

Other academic networks have been important to me. From the European Society for Ecological Economics, colleagues including Irene Ring, Clive Spash, Giogios Kallis, Sigrid Stagl, Begüm Özkaynak, Tommaso Luzzati, Ines Omann and Kate Farrell. From the European Association for Evolutionary Political Economy, colleagues including Irene Monasterolo, Andrea Roventini, Stefano Battiston, Nick Silver, Yannis Dafermos, Maria Nikolaidi and Eric Kemp-Benedict. I have also benefited from interactions with colleagues attending a series of workshops on exergy economics that we have held at Leeds and Sussex, including Bob Ayres, Benjamin Warr, Reiner Kümmel, Tiago Domingos, Tânia Sousa, André Serrenho, Matt Heun and Jonathan Cullen.

I have learned a lot from the researchers and current and former PhD students that I have been fortunate to work with, including Steve Hall, Frin Bale, Nick McMullen, Katy Roelich, Noam Bergman, Jonathan Busch, Stathis Arapostathis, Till Stenzel, Nicoletta Marigo, Ronan Bolton, Matthew Hannon, Dan O'Neill, Marco Sakai, Sam Pickard, Hannah James, Ray Edmunds, Paul Brockway, Rici Marshall, Joanne Robinson, Elke Pirgmaier, Liz Morgan, Lina Brand-Correa, Lukas Hardt, Claire Copeland, Jack Miller, Bryony Parrish and Chantal Naidoo.

Special thanks to Michael Grubb, who interrupted his family holiday to provide very useful feedback at the discussion session, Jeroen van den Bergh, who hosted a seminar that I gave at the Autonomous University of Barcelona on some of the ideas in the book, and Kathleen Araujo, for reading and providing insightful comments on draft chapters; and to Carlota Perez, for enlightening conversations during our evening walks.

Particularly special thanks to my intellectual mentor, Peter Pearson, who has provided constant support and encouragement not only for this book, but throughout my research career.

Finally, thanks to the editorial and production team at Routledge, including Annabelle Harris, Margaret Farrelly and Katie Finn, for helping to drive the project forward and produce the final manuscript.

As this long list shows, academic research is an international collaborative effort with colleagues who span a wide range of perspectives and countries. Long may this continue!

PART I

Key issues

1

INTRODUCTION – CHALLENGES OF CLIMATE CHANGE AND ECONOMIC GROWTH

Global challenges

We tend to notice energy most by its absence: if there is a power cut and the lights go out; if our car runs of fuel; or if we can't afford to heat our homes. The availability of energy and the services that it provides are vital to the functioning of the modern world and, as this book argues, in helping to drive the growth of goods and services that we have seen in Western countries and increasingly in emerging economies. However, this expansion in the use of energy comes with increasing social and environmental costs, as well as benefits. Local air pollution from coal-fired power stations is still a severe problem in many countries, as the smogs over Beijing and other Chinese cities show. At the global scale, energy use is still closely linked to the emission of carbon dioxide (CO_2) and other gases contributing to the greenhouse effect – the trapping of heat in the atmosphere, which is causing increases in global surface temperatures and associated climatic changes. This results from the burning of coal, oil and natural gas to provide those services of power, mobility, heating and lighting, which has risen exponentially since the industrial revolution began in Britain in the eighteenth century.

At Paris in December 2015, representatives of over 190 nations reached a historical agreement to moderate the risks and impacts of climate change, by reducing emissions of these greenhouse gases. This aims to keep a global temperature rise this century to well below 2 degrees Celsius above pre-industrial levels. In order to achieve this target, a transformation in global energy systems is needed, in order to provide the services we need in more efficient ways and with renewable (including solar, wind, biomass, geothermal, wave and tidal) and other low carbon (including nuclear and applying carbon sequestration to coal and gas) sources of energy. This raises the challenge of whether this transformation is consistent with continuing high levels of economic growth in industrialised and developing economies. In order to be able to address this challenge, we need to better understand the role of high carbon energy sources in previous long waves of economic growth, going back to the industrial revolution.

At the same time, the world is still suffering the effects of the financial crisis that began in 2008. Many industrialised countries are struggling to regain rates of economic growth that they had before the crisis, despite record low interest rates and the infusion of money into the economy by central banks, known as quantitative easing. Rates of income and wealth inequality within industrialised countries are also rising back to levels last seen in the 1930s and, of course, huge disparities in equality still exist between industrialised and developing countries. Furthermore, some economists have suggested that rates of innovation driving economic growth may be slowing, raising the prospect of secular stagnation. So, for understanding the potential for ensuring future social and economic progress across the world, it is important to understand the role that energy has played in delivering prosperity up to now, and whether a transformation to low carbon energy sources means that we need a new pathway to prosperity.

Understanding energy and economic growth

In this book, we chart the historical links between increasing use of fossil fuel (coal, oil and natural gas) energy sources and growth in economic output, and the benefits and costs that this has driven. The growth in economic output in every country is usually measured every quarter (three months), and reported in news bulletins and newspaper headlines, by the rate of growth in GDP (gross domestic product). This adds up the value added in all economic activities in that country, including manufacturing, services such as commerce and retailing, and public services such as education and infrastructure spending, to give the value of the total economic output that quarter. This is considered to be politically important for two main reasons. First, the total value of economic output is perceived to be closely linked to job creation in the economy – hence, politicians' mantra of increasing 'jobs and growth'. Second, the total value of economic output is closely linked to tax revenues coming in to the government, so that higher growth enables increasing government spending on public services, without having to increase the levels of taxation. However, some economists and activists have begun to question this political focus on GDP growth. They argue that, particularly in richer countries, GDP is a poor measure of overall prosperity, as it neglects important contributors to social well-being, such as unpaid caring for children or elderly relatives, and it fails to account for negative impacts, such as environmental pollution.

So, our story needs to also look in more detail at how understanding has developed of what drives growth in national economies. What will be surprising is that, despite its centrality in many economic and political debates, economists don't really have a good understanding of what drives economic growth. It may seem obvious that the availability of high quality and relatively cheap energy sources in the form of fossil fuels, which enabled machines to substitute for human labour, has been an important driver of economic growth. As we shall see, primary energy use is closely correlated with GDP growth in industrialised countries. This view is not shared by mainstream economists, though, who do not explicitly include energy as an input to their models of economic growth. So, in this book, we need to draw on the work of a range of other economic thinkers, as well as historians and physical scientists, in order to better understand the historical relations between energy and economic growth, and implications of this for a low carbon transition.

This is important because this close historical link between energy and economic growth, and the political imperative for GDP growth, could help to explain why efforts to reduce global carbon emissions have so far had so little effect. At the moment, despite significant reductions in the cost of solar panels and wind turbines in recent years, technologies like these that use lower carbon sources of energy are still currently more expensive in most cases than technologies that use fossil fuel energy. Many mainstream economists and politicians worry that if economies switch too rapidly to low carbon energy sources then this would have a detrimental effect on national economic output, and so on jobs and tax revenues (despite the mainstream economic view that energy is not important for economic growth in the long run.)

In order to gain a better understanding of the relations between the dependence of economic growth on different sources of energy and other drivers of economic growth, including improvements in knowledge and social institutions, this book argues that we need to draw on two streams of economic thinking that are on the edge of, or outside, current mainstream economics. First, thinking about how economies rely on, or are embedded in, ecological systems that provide a flow of energy and materials and assimilate the wastes produced by economies. Second, thinking about how economies evolve over time, through the interactions between changes in systems including technologies, institutional rules and business strategies. We would suggest that bringing these two strands of thinking together could provide useful insights, as we hope to begin to show in this book.

Dependence of economies on ecological systems

Different perspectives have developed on the dependence of economies on natural ecological systems, with debates intensifying in recent years, stimulated by concerns on the scarcity of resources and the severity of environmental impacts, including climate change. There is now a range of evidence that if we do little or nothing to transform economies to use low carbon energy sources, then the social and economic consequences of unabated climate change would be severe or even catastrophic, in terms of extreme weather events, such as droughts, floods and storms, rising sea levels that could engulf coastal cities, and the knock-on social consequences of famines, disease and enforced migration.[1] However, differing views have emerged of what degree of change in economic systems is required to achieve a low carbon transition.

Environmental economists apply the tools of mainstream economics, such as market incentives, to take account of environmental dependence of economies.

This leads some to argue for the potential economic benefits of this transition. They contend that the high levels of public and private investment needed to transform economies to use low carbon sources of energy, together with other related smart and efficient technologies needed, could drive a new wave of 'green growth'. This, they say, would create new jobs to replace those in fossil fuel intensive industries, and bring in tax revenues to replace those lost from these industries.[2] Moreover, they argue that, without continuing high levels of economic growth during this transition, it would be much harder to generate the levels of investment needed for the transition.[3]

Ecological economists start from the position that we should understand economies as subsystems of natural systems, which supply energy and resources to economies and receive the wastes produced, including CO_2 and other greenhouse gases. They also point to the critique of GDP growth as a measure of social well-being, and the lack of attention paid to rising levels of social inequality. This tends to lead to more radical prescriptions of changes needed in economic systems to incorporate these ecological dependencies and social factors. They argue that this requires a more fundamental reassessment of how we understand and measure the value of economic activities. This leads some to argue that we need to pursue an approach of generating 'prosperity without growth', i.e. that a transition to low carbon energy sources requires a transformation in how we measure economic value, and how we see the roles of public and private economic activity.[4] This would redefine how we measure GDP growth or seek to do away with it as the overriding political objective of economic management.

A third group of biophysical economists point to a further challenge of reconciling economic output and energy use. They argue that, despite the new extraction of shale gas in the US, the high energy surplus provided by fossil fuels over the recent centuries may be drawing to a close. They argue that this high energy surplus, i.e. the difference between the energy content of the coal, oil or gas extracted from a mine or reservoir and the energy needed to be harnessed to extract it, is a significant driver of economic growth.[5] As extraction of fossil fuels needs to go to more inaccessible places and regions, such as oil from deep sea or Arctic regions, or less concentred sources, such as shale gas, this requires higher levels of energy input per unit of energy output, so reducing the energy surplus provided. They also argue that renewable sources of energy may also provide a lower energy surplus, since they are naturally less concentrated sources, though this depends on the physical location as well as the conversion technologies. If it is the case that the energy surplus of both fossil fuel and renewable energy sources is reducing, they argue that more of the economic output of national economies will be required to be diverted to invest in the harnessing of energy, meaning that less economic output will be available for 'discretionary' spending, i.e. spending on wants rather than needs, so reducing rates of economic growth.

The book will draw on these different strands of thinking, and highlight how they give rise to different, and sometimes even contradictory, implications for economic policy.

Evolution of economic systems

How economies change over time has been a central question of economic thinking, dating back at least to the differing conceptions of the drivers and benefits of economic change proposed by Adam Smith in the eighteenth century and Karl Marx in the nineteenth century. Central to this has been debate about the role of markets that provide a structure for economic transactions, and governments that provide the social and institutional context within which markets work, usually with the stated aim of improving the wellbeing of people. So, without being able to do justice to all these theories, the book will aim to highlight how these ideas have informed how change in economies happens.

One perceptive thinker that we will draw on is the twentieth-century Austro-Hungarian economist Karl Polanyi, who we would suggest has been unjustly neglected by mainstream economists. His 1944 book, *The Great Transformation*, described the evolution of market societies.[6] Polanyi recognised the role of market economies in promoting economic development through enhancing trade and technological innovation. However, he argued that markets are never completely 'free', but are embedded in societies. Unrestrained markets would tend to undermine the social relations that underpin coherent societies. Crucially, he argued that governments had a key role in creating new institutions, such as welfare and unemployment benefits, to mitigate these negative consequences, in order to reduce social inequality and help to maintain coherent societies.

More directly, we will apply the ideas of another branch of economic thinkers, known as evolutionary economists, who study how economies evolve over time. This approach looks at processes of innovation and dynamic interactions between technological change and change in other elements of economic systems. In particular, some evolutionary economists have examined how these interactions have led to long waves or surges of economic growth, starting with the industrial revolution in Britain in the eighteenth century.[7] Though this work identifies key roles in these surges for energy technologies, such as the steam engine and electricity, it has largely developed independently of the thinking on the ecological dependence of economies. This book seeks to bring these two strands of alternative economic thinking closer together.

In particular, we shall examine the role of energy in these previous long waves or surges of economic growth. We shall see that the use of new energy sources and their efficient conversion to provide useful work and energy services have been key to these surges of economic growth. This historical analysis is underpinned by a view of economies evolving through interactions between technologies, institutional rules, strategies of businesses and practices associated with energy use, with dependencies on flows to and from ecosystems.

This will lead us to look at recent ideas from ecological economists relating energy provision to economic growth, and insights from economic historians on these relations. This enables us to shed light on recent debates as to whether a low carbon energy transformation could lead to a new surge of 'green growth' or whether it would require a more fundamental shift in economic priorities. Finally, we shall draw these strands together to identify elements of a new pathway to prosperity for a sustainable and low carbon future.

This book does not aim to provide all the answers to the question of future paths of energy use and economic development. However, by providing a historical understanding of the relevant connections between physical, social and economic changes, we hope to enable the reader to be able to understand better the link between their own energy use and global economic and environmental systems, and to be able to ask the right questions of our political and business leaders.

Structure of the book

We now set out the structure of the arguments in the book. In order to better understand the economic implications of energy use, and to be able to compare different energy sources and technologies, we need to have a basic grasp of what is energy and why it is important. Unfortunately, the range of energy units and how they are used can be confusing, so Chapter 2 will guide the reader through these. We then move on to outline three challenges relating to the future role of energy in relation to economic development. First, the clear evidence linking climate change to human-induced emissions of greenhouse gases, predominantly from energy use, requires a transformation in the sources and ways that we use energy in all countries. The higher per person rates of energy use and historical responsibility for emissions creates a duty on Western industrialised countries to take a leading role in this transition. Many countries, including the UK and Germany, have made strong long-term commitments to an energy transition, and have started to take action to promote the further development and take-up of low carbon energy technologies. Second, around 2 billion people in developing countries do not have access to high quality energy sources. Those fortunate enough to live in industrialised countries already use a large amount of energy per person, some more efficiently than others. Most of the growth in energy use in the coming decades will come from the industrialisation and spread of energy use in emerging and developing economies. Third, and finally, the declining rates of net energy return on the production of fossil fuels, as production is forced to move to less accessible sources in deep water and Arctic regions. This could have severe economic implications as more and more economic output has to be diverted to investment in producing energy.

Chapter 3 sets out perspectives on the ecological dependence of economies, and on evolutionary theories of economic change. This leads to the coevolutionary framework that we use to examine the role of energy in industrial change over the long term. This shows that changes in the availability and uses of energy sources have led to wider changes in social, economic and environmental systems, but that social, economic and environmental factors have also strongly influenced choices of energy sources. This is investigated by examining mutual changes in technologies, institutional rules, business strategies, practices of energy use and natural ecosystems.

Chapters 4 to 7 use this framework to trace out the evolution of ever-expanding human use of energy. In general, this has been closely related to the expansion of the scale of economic activity and improvements in human wellbeing, such as increases in healthy life expectancy. In Chapter 4, we examine pre-industrial energy systems. These were mainly human and animal labour, which rely on solar energy embodied in food, and small amounts of energy from wind and water flows. As a result, there was little excess energy beyond that required for basic food and material needs. Chapter 5 examines the first industrial revolution, which began in the UK in the eighteenth century. Historians have argued that access to higher quality forms of energy, first water power and then steam power from the burning of coal, enabled the start of a positive feedback loop. The substitution of human and animal labour by energy-using machines enabled more efficient production and transportation of goods, which stimulated demand and enabled more efficient extraction and conversion of energy sources. This is a key engine of the sustained growth in the size of national economies, which began at this time.

Chapter 6 traces two subsequent 'long waves' of energy-industrial change, which still form the basis of modern Western economic activity. First, that associated with the rise of electrification in the late nineteenth and early twentieth centuries. The discoveries and inventions by Michael Faraday, James Clerk Maxwell, Nikola Tesla, Thomas Edison and others showed that a moving magnet could generate an electric current, and a changing electric current could turn a magnet, thus converting between energy of motion and electrical energy. This formed the basis for the electric generator and electric motor that would revolutionise production systems, such as automated assembly lines, and subsequently household energy-using devices and practices. Second, the use of oil and natural gas enabled rapid changes in transportation and heating services in the twentieth century. In particular, oil and the fuels resulting from its refinement, including gasoline (petrol) for motor vehicles, diesel fuel for trucks and trains, and kerosene for aircraft, enabled both mass transportation of goods and mass movement of people for business and leisure reasons.

Chapter 7 traces how these developments have enabled the rise of the consumer society. Changes in practices of energy use, such as enhanced leisure and retail activities, have coevolved with changes in technologies linked to electrification and oil, and changes in institutional rules and business strategies. These have developed to take advantage of the possibilities enabled by relatively cheap, high quality energy sources and conversion processes. This resulted in a new spurt of economic growth in Western countries in the 1950s and 1960s. Of course, enhanced energy use was not the sole driver of change, and other developments in social rules and norms, such as limited liability companies and weakening of class barriers, were important. Nevertheless, we argue that having access to affordable energy services was a key enabler of change. Even though some have argued that the rise of information technologies has the potential to deliver a new wave of economic growth, it is important to remember that all those smart phones, tablets and laptops that we use, and the massive information processing systems that they are connected to, still use large amounts of electrical power.

Though these historical cases illustrate the important role of energy transitions in economic development, mainstream understanding of economic growth does not include a role for energy inputs. Chapter 8 examines recent work showing that the availability and efficient conversion of energy sources has been a key driver of economic growth. In particular, electricity is a high quality energy source that is able to be used for many functions, and can be generated from many primary sources. Together with the efficient conversion of the chemical energy in oil to provide motion via the internal combustion engine, these high quality energy sources helped to drive the twenty-five fold increase in world economic output that has occurred since 1900.

Chapter 9 provides a range of perspectives from economic historians on relations between energy use and economic growth and from researchers examining declining net energy returns. This leads us to question whether low carbon energy technologies yet have the properties to drive a new industrial revolution, similar to those of past economic surges.

Chapter 10 considers potential future energy pathways that could reconcile increasing access to high quality energy whilst significantly reducing human-induced carbon emissions and taking account of declining rates of net energy return. These are likely to require major changes in the way that energy systems are governed and in the practices for which energy is used, as well as radical technological changes. These system changes are likely to be at least as dramatic and wide-reaching as the previous energy-industrial system changes that we have examined in previous chapters. We examine some of the issues relating to realising these pathways, including the likelihood of 'stranded assets' in high carbon energy systems that cannot be used, how to avoid 'rebound effects' in which increasing energy consumption takes back savings from energy efficiency improvements, and the need to reorient financial systems to deliver the investment needed. Finally, we consider the potential for more decentralised energy pathways, in which citizens play a greater role in their own energy provision.

Chapter 11 steps back to look at these challenges in the context of recent debates about the desirability of continuing the present focus on achieving high rates of economic growth. Some argue that economic growth is no longer delivering widespread prosperity or happiness in industrialised countries, as more and more of us are having to work longer in order to be able to buy more and more material goods and services that do not increase our wellbeing. Others argue that 'green growth' is necessary to deliver the high levels of investment needed for a low carbon transition. Recent thinking argues for a perspective based on assessing options in terms of their impact on a range of indicators of human wellbeing, rather than purely on economic growth.

Despite the scale of these challenges, Chapter 12 seeks to end on a hopeful note. The fact that we can identify the challenges and map out pathways to addressing them in itself suggests that achieving an energy transition whilst maintaining a prosperous economy is not an impossible task. It will almost certainly require greater changes to our current consumption practices than has yet been acknowledged by governments, businesses or most of the public. But then failure is not an option. If we continue on our current energy-profligate pathway, only making incremental changes and hoping that some new miracle technological solution will save us from hard choices, then the social and economic consequences of a world facing dangerous climate change and declining net energy sources are unthinkable. In the short term, at least, investment in low carbon energy and related resource-efficient technologies could help to drive a new surge of 'green growth'. However, in the longer term, more radical changes to current economic priorities may be needed to realise a sustainable and low carbon energy system transformation. We conclude by identifying key elements of a new pathway to prosperity.

Notes

1 The likely physical impacts of climate change are spelled out in reports by the Inter-governmental Panel on Climate Change (IPCC), see IPCC (2014). The potential economic impacts were detailed in the Stern Review (2007). Accessible popular accounts of climate change and how can it be addressed include Berners-Lee and Clark (2013); Klein (2015); Stern (2015).
2 See, for example, the New Climate Economy report of the Global Commission on the Economy and Climate (2014).
3 See Hepburn and Bowen (2013).
4 See Jackson (2009/2017).
5 See Hall and Klitgaard (2012).
6 Polanyi (1944/2001).
7 Freeman and Louçã (2001); Perez (2002).

2

WHAT IS ENERGY AND WHY IS IT IMPORTANT FOR THE ECONOMY?

What is energy?

Energy in the forms of electric power, heat, light and motive power is vital to both the production and consumption of goods and services by businesses and households. But what is energy, and why is it such a useful idea? Our modern understanding of energy developed in the eighteenth and nineteenth centuries, in parallel with the important developments in the use of machines to provide power that we will discuss in later chapters. However, it is important to realise that this was not a linear transfer of ideas from basic scientific understanding to applied knowledge. In some cases, industrial pioneers such as Richard Trevithick and James Watt developed steam engines in advance of a full theoretical grasp of the underlying concepts. In other cases, such as the advances in understanding of the principles of electricity and magnetism by Michael Faraday and others, this did lead directly to new inventions such as electric motors, dynamos and telegraph communications.

The key insights developed by James Prescott Joule, Lord Kelvin (William Thomson) and others in the nineteenth century were that different forms of energy could be converted into one another, and that heat can be understood as a transfer of energy from one system to another. In 1843, Joule showed that a falling weight can turn a paddle in a container of water and that the resulting motion of the paddle will add energy to the water, raising its temperature. The key point is that the energy of motion of the paddle is converted into the internal energy of the water, in the form of its higher temperature. Energy is transferred to the water by the mechanical work done by the rotating paddle. Energy can also be transferred to a system directly in the form of heat by connecting a hotter body to a cooler body. Heat will then flow, reducing the internal energy of the hotter body and increasing the internal energy of a cooler body.

This leads us to a definition of what energy is:

Energy is the capacity to do work or transfer heat.

Energy comes in many forms. The chemical energy stored in petrol (gasoline) is harnessed to provide the mechanical work of driving a car forward. The energy of motion of a magnet in a coil of wires is converted to electric energy in the wires, which can then be converted to light energy or to power the electrical circuits in a computer. The burning of natural gas can be used to provide hot water and space heating by converting chemical energy into heat energy.

Crucially, Joule showed that if you add up all the energy before a conversion, then the same amount of energy is present after the conversion process. This became known as the First Law of Thermodynamics:

Energy can be converted, but never created or destroyed.

This being the case, strictly speaking, human activity does not create energy, but relies on ingenious ways of harnessing natural flows of energy. The main sources of energy for doing work that we use on planet Earth are chemical energy in fossil fuels, (direct and indirect) solar energy, tidal energy, geothermal energy and nuclear energy. Solar energy is by far the largest available source, delivering energy to the Earth at a continuous rate of 174 million gigawatts (GW). (Solar energy derives from the energy released in fusion of hydrogen into helium in the hot, dense core of the sun, according to Einstein's famous equation $E = mc^2$.) Green plants harness this solar energy by the process of photosynthesis, in which, with the help of carbon dioxide, some of the incident solar energy is converted to energy stored in chemical bonds. The plants use this chemical energy to power their metabolic processes of survival, growth and reproduction. When animals eat plants, they are able to convert this stored chemical energy into internal chemical energy sources to power their own metabolic processes of survival, growth and reproduction, with the help of oxygen breathed in from the air. As the solar energy that ultimately powers these processes will continue to arrive on the Earth at roughly the same rate for as long as the Sun continues to shine (around another 5 billion years), solar energy is considered the main form of renewable energy. Solar energy can be directly converted to electricity, using photovoltaic (PV) panels, or used to heat water in homes. Concentrated solar power stations use solar energy, concentrated by large mirrors, to heat water to produce steam to drive a turbine, which generates electricity. This has the advantage that the solar energy can be stored for up to 12 hours in the form of hot molten sand, before being used to generate electricity.

Fossil fuels (coal, oil and natural gas) are the result of the stored chemical energy from sunlight in plants and animals that lived millions of years ago. As these were buried and crushed by overlying rock, this chemical energy is converted to more dense energy sources by natural heat and pressure. This more dense chemical energy in the fuels can then be converted to energy of motion, heat energy or electrical energy by being burned in oxygen in engines, boilers or power stations. In a conventional power station, coal or natural gas is burned to heat water into steam to drive a turbine to generate electrical energy, releasing carbon dioxide into the atmosphere. In addition to the environmental impacts of burning these fuels, fossil fuels are not considered renewable because we are now using them at a much faster rate than they were created. This leads to concerns that the supply of these fuels, particularly oil, may soon hit a peak level.

Other forms of renewable energy, such as biomass, wind and wave power, are indirect forms of solar energy. Biomass or biofuels use the solar energy stored in plants to provide heat, electrical energy or motive power. As this is less concentrated than fossil fuel energy sources, it provides a lower energy surplus. Wind and wave power rely on the fact that solar energy heats different parts of the Earth to different degrees, leading to flows of wind in air and waves in water, which can then be converted into mechanical or electrical energy.

Tidal energy is the result of gravitational interactions between the Earth, Sun and the Moon. The gravitational pull of the Moon on sea water on the Earth's surface raises two tides per day. The energy released in tidal flows can be converted into electric energy. You might ask what happens to this energy if it is not converted into electric energy? In this case, it is dissipated as heat which slightly raises the temperature of the water. As we shall see below, this is a less useful form of energy.

Geothermal energy is the harnessing of the natural heat energy under the surface of the Earth. This results from the initial formation of the Earth, as well as heat released by the radioactive decay of naturally occurring elements.

Energy can also be released by the nuclear reactions between atoms. There are two forms of nuclear energy. First, nuclear fusion is the process of release of energy by the fusion of light elements. As noted, the fusion of hydrogen (the lightest element) into helium in the heart of the Sun is the process that generates solar energy. Second, nuclear fission is the process of release of energy by the fission (splitting) of the atomic nucleus of heavy elements, such as uranium. The splitting of the atomic nucleus typically also releases particles called neutrons, which can trigger fission of further heavy atoms in a chain reaction. A controlled nuclear (fission) chain reaction will release heat energy, which can be used to heat water into steam to drive a turbine to generate electrical energy, as is done in a nuclear power station. An uncontrolled nuclear fission chain reaction leads to a meltdown or a nuclear bomb.

These forms of energy are referred to as 'primary energy' in the context of providing inputs into energy conversion processes for human activities. Forms of energy after conversion for human use, such as refined petroleum or electrical energy, are referred to as 'secondary energy'. These typically require a further conversion process in end-use devices, such as vehicles or electrical devices, which convert this 'secondary energy' into 'useful energy' or 'useful work'. This 'useful work' then enables the provision of the services that we want, such as mobility, working devices, warmth and illumination (see Figure 2.1).

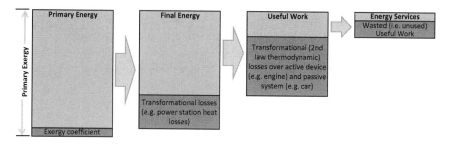

FIGURE 2.1 Energy chain diagram.
Source: Adapted from Brockway et al. (2015).

How do we use energy?

However, this raises a challenge for our everyday understanding of energy, where we talk about 'energy shortages' or 'running out of energy supplies'. If energy is never destroyed, then why can't we just convert some more into useful sources that we can use? It turns out that some energy sources are more useful than others. Experiments by French physical scientist Sadi Carnot in the early nineteenth century studying steam engines showed that, in any conversion process, some energy is converted to 'useless' heat, which cannot perform any further useful work. This can be understood as a process of conversion of ordered energy into disordered energy. In this process, the total amount of disorder, measured by a quantity called 'entropy', increases. This is known as the Second Law of Thermodynamics:

> In any energy conversion process, entropy increases and the amount of useful energy available to do work decreases.

The easiest way to think about this is that there are many more ways to be disordered than to be ordered, so, in any natural process, the amount of disorder is greater at the end of the process than at the start. For example, putting a hot gas together with a cold gas will lead to them equalising their temperature at an average value. This conversion process could be used to drive a turbine and produce work. However, at the end of the process, we would be left with a gas at a uniform temperature, which can do no further work.

The capacity of a physical system to do useful work is measured by a quantity called 'exergy'. This combines insights from the First and Second Laws of Thermodynamics.[1] Fortunately, solar energy is a high exergy energy source, which means that most of its energy is available to do work. Fossil fuels, such as coal, oil and natural gas, are also high exergy energy sources.

Why is this important to our story? Because any living system is subject to decay through increasing entropy, it requires a flow of ordered (high exergy, low entropy) energy to grow and maintain its systems, so that it can survive and reproduce. Indeed, this is one definition of what is life.[2] The ordered solar energy coming from the Sun is the basis of life on Earth, through the process of photosynthesis in plants. This does not contravene the Second Law of Thermodynamics, because living systems expel high entropy waste to their environment. We shall see how this process of a flow of energy, involving low entropy being dissipated to high entropy, used to power and maintain systems also applies to human-produced systems.

So, an energy conversion process, such as a power station, involves taking a concentrated, ordered form of energy, such as a flow of solar energy or the stored chemical energy in a fossil fuel, and converting it into another form of energy, such as electrical energy, which can be used to do work, such as power an electrical device. A key consideration is the efficiency of any conversion process. The most widely used definition of energy efficiency refers to the First Law above:

Energy efficiency = useful energy out/energy input

In a coal-fired power station, coal is burned at high temperatures to heat water into high pressure steam that turns a turbine to generate electricity. Only about 35–40 per cent of the energy content of the coal is converted to electrical energy. The rest of the energy ends up as high entropy energy in the heated water, which has to be cooled to start the cycle again. The energy content of this heated water depends on its temperature, which is typically only around 20°C higher than that of water in the environment, after it has transferred energy to drive the turbine. So, its capacity for doing useful work (i.e. its exergy) is low, and it is not usually considered as part of the useful energy output. Other things being equal, improving the energy efficiency of a process reduces the amount of energy input needed to produce a given amount of useful energy output. However, as we discuss in more detail later, if this leads to an increase in the amount of demand for that useful energy, the so-called 'rebound effect', then energy input may be reduced by a smaller amount or even increased.

Though the first law energy efficiency definition is appropriate for considering a single energy conversion process, we often want to compare different processes for performing a given task. It then makes more sense to consider the exergy efficiency of a conversion process:

Exergy efficiency = useful work out/exergy input in
= minimum primary exergy input required to
perform a task/exergy input in

This then measures how far a process is from its theoretical maximum efficiency. For, if the exergy input in was equal to the minimum exergy required to perform that task, then the process would have an exergy efficiency of 100 per cent, and no losses would occur. In practice, tasks are usually performed with more than the minimum primary exergy required, and so the process would have an exergy efficiency of less than 100 per cent, and some exergy (i.e. the potential to do useful work) is lost in the process. However, electrical energy can be converted to other forms of energy at close to 100 per cent exergy efficiency, making it a very useful form of energy.

In general in this book, when we are talking about energy, we are usually referring to exergy, as the output we want is useful work. Sometimes, we need to be clear about the difference between energy and exergy, though. Fortunately, though, energy and exergy are measured in the same units. We now turn to those units of energy measurement.

Units of energy

In order to compare different energy sources and energy conversion processes, we need to quantify energy. Unfortunately, a range of different units for amounts and flows of energy are used, which can lead to confusion.

On your household energy bill, the amount of electricity that you have used in a given period, say a year, is measured in kilowatt-hours (kWh). This unit is equivalent to a thousand watts of electrical energy being used every hour for that period. A watt (W) measures the rate of flow of energy in joules per second, where the joule (J) is the fundamental international unit of energy.[3] So,

$$1 \ kWh = 1000 \ W \times 1 \ \text{hour}$$
$$= 1000 \ J/s \times 3600 \ \text{seconds}$$
$$= 3{,}600{,}000 \ J$$

A typical UK household uses around 80 kWh of electricity per day, or 29,200 kWh per year. This is equivalent to 29.2 MWh per year, where a megawatt-hour (MWh) represents a million watt-hours.

A large coal-fired power station is typically rated to produce around 1000 million watts (MW) at maximum power output. Over a year, due to outages for maintenance and repairs, it will typically be operating at this output for only 50 per cent of the time. So, it will supply over the year

$$1000 \ MW \times 50\% \times 8760 \ \text{hours} = 4{,}380{,}000 \ MWh$$

Hence, it can supply

$$4{,}380{,}000 \ / \ 29.2 = 150{,}000 \ \text{households}$$

The important point here is that we need to be clear about whether we are talking about an amount of energy, either in a given material or over a given period of time, which is measured in joules (J) or kilowatt-hours (kWh), or a rate of flow of energy, measured in watts (W).

Of course, other energy units are available. In the UK, the energy content of natural gas is measured in therms. In the US (and formerly in the UK), electric or heat energy is measured in British thermal units (Btu). One kilowatt-hour (kWh) is equivalent to 3,416 *Btu*. For large amounts of energy, Americans use the quad, equivalent to a quadrillion (thousand million million) *Btu*.

The energy content of oil is usually measured in terms of the equivalent amount of energy contained in a (metric) tonne of oil, i.e. in tonne of oil equivalent (*toe*). The high energy density of oil means that

$$1 \ toe = 41,900,000,000 \ J$$
$$= 11,640 \ kWh$$

Note that the principle of conservation of energy (First Law) means that it is always possible to convert between these different units of energy, but different sources of energy have different characteristics in terms of their usefulness, e.g. in relation to practical rates of energy conversion.

Comparing energy sources and uses

As we noted above, useful energy is lost every time energy is converted from one form to another. So, in comparing different energy sources and uses, it is important to be clear whether we are talking about (primary) energy inputs into the economy or amounts of useful energy being used in the economy. In terms of impact on the environment – either resource consumption or greenhouse gas emissions – it is the primary energy inputs that matter. In terms of contribution to economic activity and human welfare, it is energy use that matters. As we shall see, availability of energy sources, efficiency of conversion of primary energy into useful work, and structural changes in the mix of energy sources and the types of end uses have all been important aspects of past energy transitions.

It is also important to be clear whether we are talking about amounts of energy consumed or rates of consumption, i.e. power. For the most part, we use kilowatt-hours (*kWh*) as the basic unit of amount of energy. When we go up to country scale, then we use terawatt-hours (*TWh*), which is a billion kilowatt-hours. Similarly, rates of energy flow are measured in watts (*W*), million watts (*MW*) or billion watts (*GW*). Though primary energy values are often given in million tonnes of oil equivalent, we will present these in terms of terawatt-hours (*TWh*).

According to the International Energy Agency (IEA), in 2014, the total primary energy supply into the world economy was just under 160,000 *TWh*.[4] Thanks to growth of energy consumption in both industrialised and emerging economies, this was double the world's total primary energy supply in 1973. Fossil fuels made up the majority of this energy supply in 2010, with oil making up 31 per cent of the total, coal 29 per cent and natural gas 21 per cent. Nuclear power provided 5 per cent of the total, hydro power 2.5 per cent and biomass (crops, residues and waste) 10 per cent. Other renewables, including solar, wind and tidal power, provided around 1.5 per cent.

From this supply, the world used 110,000 *TWh* (and so around a third of the energy content of primary sources is lost, as waste heat and other conversion losses). Just over half of this was used by the richer countries of North America, Europe, Australasia and the former Soviet Union, despite having less than a quarter of the world's population. Of the other countries, the fastest growth in energy consumption over the last 30 years has come from China, other emerging Asian economies and the energy-rich Middle Eastern countries. The largest end use is transport of people and goods, mainly using oil, followed by heating, cooking and electric power, which makes up about a fifth of final energy use.

Carbon emissions

As we shall explore in later chapters, this level of energy consumption in industrialised countries is far above that of pre-industrialised societies, and has contributed significantly to growth in the size of economies and improvements in human welfare. However, the heavy dependence on fossil fuel sources, together with rapid growth in world energy consumption, creates huge challenges for maintaining and enhancing global prosperity whilst significantly reducing carbon emissions. In 2014, the world emitted just over 32,000 million tonnes of carbon dioxide from fuel combustion (46 per cent from coal, 34 per cent from oil and 20 per cent from natural gas). This is about 16 times larger than carbon emissions in 1900, and more than double the emissions in 1973.

The scientific evidence, collated by the Intergovernmental Panel on Climate Change (IPCC), clearly states that warming of the global climate system is unequivocal, and that this is very likely (greater than 90 per cent chance) due to human emissions of greenhouse gases, mainly from energy supply and land use changes, such as destruction of rainforests. However, despite the signing of the UN Framework Convention on Climate Change in 1992 and subsequent Kyoto Protocol, which committed industrialised countries to reduce their emissions by an average of 5 per cent by 2012 from a 1990 baseline, emissions are still growing at around 2 per cent per year. What matters from the perspective of the climate is the cumulative emissions over time, leading to rising concentration of greenhouse gases in the atmosphere. Concentrations of CO_2 have been rising in the atmosphere since the industrial revolution as a by-product of the massive increase in burning of coal, oil and natural gas to provide energy and the clearing of forests and other vegetation that absorbs CO_2. As visible and ultraviolet light from the Sun reach the Earth with an irradiance of 1050 watts per square metre (W/m^2), this heats the surface of the Earth and a similar amount of infrared radiation is emitted by the Earth, otherwise it would overheat. The CO_2 in the atmosphere absorbs some of this infrared radiation, trapping it in the Earth's atmosphere, causing the planet to warm. This is known as the greenhouse effect. Other gases from human activities, including methane (CH_4) and nitrous oxide (N_2O), are present in the atmosphere in smaller concentrations, but they are much more potent greenhouse gases.

Building on the work of French physicists Joseph Fourier and Claude Pouillet and British physicist John Tyndall, in 1896, Swedish physical chemist Svante Arrhenius was the first to quantify the effects of increasing the concentration of

CO_2 in the atmosphere.[5] On basic physical principles, he predicted that doubling the concentration of CO_2 would lead to a global warming of 4°C. Pre-industrial concentrations of CO_2 were around 280 parts per million (ppm) and 250 years of industrial activity have led to a CO_2 concentration of 400 ppm in 2016. On current business as usual trajectories, as the world consumes more and more energy from fossil fuels, atmospheric concentrations of CO_2 are set to double by the second half of this century. The latest scientific analysis, taking into account the positive and negative feedbacks due to cloud formation, depleting reflection of sunlight by ice and the effects of other greenhouse gases, calculates that this would lead to a global average temperature rise of 3–4°C, similar to Arrhenius's original calculation. However, there are concerns that positive feedback effects could lead to higher levels of warming. For example, large amounts of methane are known to be trapped in permafrost in Arctic regions. As the planet warms, some of this permafrost will melt, releasing the trapped methane, leading to further warming. It is very difficult to estimate the likely size of these types of effects using current climate models, but a precautionary approach would suggest the need for planning for the potential for significant effects.

The concentration of carbon dioxide in the atmosphere has been directly measured by Charles Keeling and colleagues at Mauna Loa in Hawaii since 1958, when it stood at 315 ppm. In 2013, CO_2 concentration passed 400 ppm for the first time (Figure 2.2). Methane, nitrous oxide and other greenhouse gases add the equivalent of at least another 30 ppm.

The burning of coal, mainly for industry and electricity generation, and oil, mainly for transport, are the largest sources of carbon emissions from energy use. Burning natural gas emits about half the carbon emissions of coal for the same amount of energy supply. Modern renewables, including generation of electricity

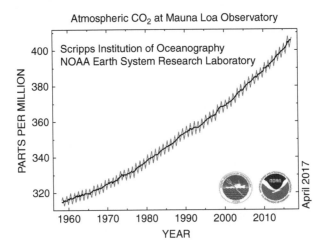

FIGURE 2.2 Atmospheric CO_2 concentrations since 1958.
Source: Scripps Institution of Oceanography/NOAA Earth System Research Laboratory.

from solar, wind, wave or tidal power, and nuclear power stations emit low levels of CO_2 per unit of energy generated. For a full accounting for all sources, it is necessary to take into account the emissions associated with extraction and transmission of fuels, as well as their final use. For most energy sources, this adds only a small percentage to their emissions over the full life cycle, but this could be significant if we need to reduce carbon emissions to very low levels. When it comes to producing energy from growing crops, wood and other residues, usually referred to as biomass, it is essential to consider the life cycle emissions. Biomass energy is taken to be renewable, as the carbon emissions resulting from their conversion to heat or electricity should be offset by the carbon emissions absorbed in the plant's growth. However, it is vital to take into account indirect effects which could undermine this balance. For example, if a rainforest or peat land is cut down to create a plantation for energy crops, then the CO_2 absorption capacity lost could take several decades to pay back. Growing energy crops can also compete with land for growing food, potentially leading to increases in food prices.

In December 2015, after previous various failed attempts, 195 countries came together to reach the Paris Agreement on mitigating human-induced climate change.[6] The signatories to the Paris Agreement agreed to act to hold the increase in the global average temperature to well below 2 °C above pre-industrial levels and to pursue efforts to limit the temperature increase to 1.5 °C above pre-industrial levels. In September 2016, President Barack Obama of the US and President Xi Jinping of China agreed to ratify the Agreement on behalf of their respective nations. The Agreement came into force in November 2016, and has been ratified by nations representing over 80 per cent of global emissions of greenhouse gases. Though the new US President Trump announced on 1 June 2017 that the US will withdraw from the Paris Agreement, President Xi has indicated that China will continue to abide by the Agreement and will take a leading role in its implementation.

Achieving the targets to mitigate climate change under the Paris Agreement will require a transformation of global energy systems to dramatically reduce and, as far as possible, eventually eliminate carbon emissions associated with energy use. However, this needs to be done whilst addressing the challenge of achieving continuing global economic development, in the face of high levels of global inequality.

Global economic development

As we shall see, five great surges of technological and economic change have delivered amazing improvements in the quality of life of most people in industrialised countries. Access to new energy sources and improvements in the efficiency of conversion to provide useful work have been crucial in these changes. This is not only directly – in terms of improvements in the quality and reductions in the costs of providing a range of energy services, including lighting and heating of homes and businesses, electric power for a huge array of end-use devices, and fuel for transport and mobility of goods and people; and indirectly – through reductions in the costs of manufacturing products and the use of new materials, such as plastics, which are derived as by-products of energy production processes; but also new energy sources and efficiency improvements have contributed to the wide range of products and services available to the average citizen of a rich country. Furthermore, modern agriculture is intensely dependent on fossil fuel sources for providing the fertilisers which help to enable the high yields of arable crops for human consumption and fodder for animal consumption that make up Western diets.

However, around 2 billion people in developing countries do not yet have access to high quality energy sources. Most of the growth in energy use in the coming decades will come from the industrialisation and spread of energy use in emerging and developing economies. This is necessary in order for those people to escape desperate poverty and to be able to live fulfilling and satisfying lives. The question is what form of development they will follow, and will they aspire to achieve the high energy consumption lifestyles that those of us in richer countries now enjoy?

The main example of economic development over recent years, that of China, is not necessarily one for other countries to emulate. Over the last 20 years, China has gone from an economy the size of Belgium's to the second largest economy in the world. This has undoubtedly helped many millions of Chinese citizens to escape from poverty, mainly by moving from subsistence agricultural lifestyles in rural areas to (low-) paid work in rapidly expanding cities. This has been enabled by a massive increase in China's energy consumption, mainly in the form of coal use for heavy industry and heating of homes and businesses. This industry has enabled the country's rapid economic development, with annual economic growth rates of over 10 per cent during the 2000s, by manufacturing products for export to richer countries. However, the social and environmental costs of this transformation have been very high. China's carbon emissions have risen dramatically, so that it is now the world's second largest emitter of greenhouse gases (though still much lower than Western countries on a per person basis). Of more immediate impact is that of local particulate and air pollutant emissions that have caused a dramatic reduction in air quality, as seen in the smogs that regularly lie over Beijing and other major cities. Local water quality has also suffered, and the damming of the Yangtze and other major rivers, to provide electrical power, has contributed to increased risks of flooding.

The Chinese government is now seeking to address these challenges, including by a massive expansion of renewable energy sources in order to stabilise its carbon emissions by 2030 and reduce them after that. However, as China begins to reba-lance its economies towards more domestic consumption, its energy use would continue to grow rapidly, if it follows the pattern of Western countries.

Global inequality

Since 1971, the world's business leaders and politicians have been meeting at the annual World Economic Forum meeting in Davos in Switzerland. The Forum is widely considered a champion of the economic model of globalisation, based on liberalised free markets, that has been dominant since the 1980s. This model underpinned high rates of economic growth until the financial crisis of 2008, but critics argued that the benefits of this growth largely flowed to the bankers and corporate executives that make up the richest 1 per cent. The household income of a middle class family in the US was lower in 2015 than it was in 1999, despite the economy having grown by 18 per cent over that period and annual growth rates averaging over 2 per cent since the economic recovery in 2010.[7] This finding is repeated across industrialised countries, as improvements in labour productivity, the value produced per hour worked, have not led to corresponding improvements in real wages for the average worker (see Figure 2.3).[8]

Each year just before the Davos meeting, the UK development charity Oxfam publishes a report on global inequality. In 2017, it argued that just eight men, including Bill Gates and Warren Buffett, own as much wealth as the 3.6 billion people making up the poorest half of the world's population.[9] Of these, nearly 800 million regularly go hungry, due to not being able to produce or buy enough food to support themselves. This represents a return to levels of inequality in wealth last seen in the 1930s, but now with much higher absolute numbers of people, over 7 billion in 2015 compared to 2 billion in 1927.

French economist Thomas Piketty has conducted a detailed review of the trends in inequality over the past century.[10] Income inequality dropped sharply in the 1940s in the US, with the share of income taken by the top 10 per cent of earners dropping from over 45 per cent to less than 35 per cent of total income. Throughout the boom period of the fifth surge from 1950 to 1980, this share remained around the same, as the benefits of economic growth were shared between increasing returns to investors and increasing real wages in the US and

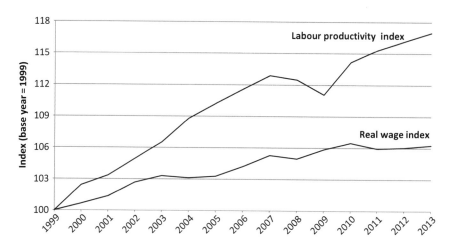

FIGURE 2.3 Decoupling of wages and GDP since 1999.
Source: ILO (2015), *Global Wage Report 2014/15*.

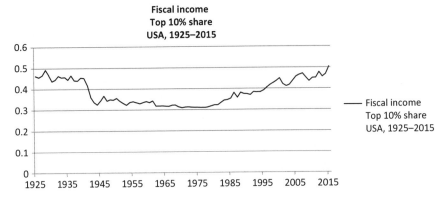

FIGURE 2.4 US income inequality, 1925–2015.
Source: World Wealth and Income Database (WID, no date) (extended from Piketty and Saez, 2003).

other industrialised countries. However, since 1980, these economies have again become more and more unequal, with the top 10 per cent of US earners now again receiving over 45 per cent of total income (see Figure 2.4). Previously, most economists had accepted the hypothesis of Belarus-born American economist Simon Kuznets that as economies developed they would initially grow more unequal but that inequality would reduce as economies became richer through industrialisation. This perspective is based on the mainstream economic view that returns to labour should be based on its contribution to productivity.

Piketty argues that the period from 1950 to 1980 was an interruption to the normal workings of capitalist economies, which naturally lead to rising inequality unless social action is taken to prevent this. He argues that, under normal circumstances, the rate of return to wealth (r), in the form of dividends, rent, interest and capital gains on investments, is greater than the rate of economic growth (g), i.e. r > g. He estimates that the rate of return to wealth will continue to be around 4–5 per cent, whereas economic growth in industrialised countries is likely to remain between 1 and 2 per cent, due to economic headwinds slowing the rate of new innovations. This implies that the already wealthy will get even wealthier, and inequality will continue to increase.

The reduction in inequality in industrialised countries between 1950 and 1980 can be traced to action by the state to create social welfare programmes, such as health and unemployment benefits, and the ability of workers, through the power of trade unions, to negotiate fair wage increases. Arguably, as we shall see, these can be linked to the general rise in prosperity linked to the deployment of the fifth techno-economic surge since the industrial revolution. In that surge, the majority of the jobs being created were in manufacturing industries with concentrated skilled and semi-skilled workforces, which supported the rise of powerful unions.

Piketty's policy proposals for responding to this inequality, by introducing a sharply rising wealth tax or progressive income tax of up to 80 per cent for the highest earners, have naturally proved very controversial. The central question is whether such sharply progressive taxes would themselves contribute to slowing economic growth by reducing entrepreneurial and wealth-generating activities. The experience of the period from 1950 to 1980 suggests that this is not necessarily the case. If economic growth is more driven by surges in technological and institutional change, as we shall argue, then the policy focus should be on promoting innovation in socially desirable directions, and this can be consistent with reducing inequality if appropriate policies are put in place to ensure that the benefits of growth are widely shared. However, as the richer members of society capture political and social power, then it becomes more and more difficult to implement these types of policies.

Impacts of inequality

Inequality has serious negative impacts on the life chances of those at the bottom of the income scale. The benefits of better education and health care flowing mostly to higher earners means that life expectancy differs significantly between higher earners and lower earners. Recent analysis by the Brookings Institute shows differences in life expectancy in the US of 12 years for middle-aged men and 10 years for middle-aged women. A middle-aged man born in 1940 in the lowest 10 per cent of income could hope to live to the age of 76, whereas a similar man in the highest 10 per cent of income could expect to live to the age of 88.[11]

Inequality may not be just impacting the poorest in society, though. British social researchers Kate Pickett and Richard Wilkinson have gathered evidence that inequality has negative social, economic and environmental consequences throughout society.[12] They show that societies that are more unequal have worse outcomes on a range of measures of social wellbeing, such as rates of mental illness, obesity, violence and welfare of children. They argue that social and psychological factors may be as important as physical factors in explaining these outcomes. Support from this comes from the so-called Whitehall studies of British civil servants from the 1960s to the 1990s. These found that, despite having above average incomes, lower-ranking civil servants had higher rates of stress-related illnesses, such as heart disease, and that this was associated with feelings of lack of control over work demands.[13] Inequality can also have economic impacts through leading to more frequent and severe boom-and-bust cycles. Finally, inequality can exacerbate environmental problems by stimulating status-driven consumption, as people seek to maintain their social status through purchases of higher cost goods and services in order to 'keep up with the Joneses'.

Through the Equality Trust that they set up, Pickett and Wilkinson have also argued for policies to reduce inequality. These could include measures to set limits between the top and bottom earners in any company, as well as regulations to ensure a living wage for all employees.

Though these findings primarily relate to inequality within countries, as the world becomes more interconnected, the challenges associated with global inequality become stark. Access to affordable, reliable and modern energy services for all is rightly one of the UN's Sustainable Development Goals, aiming to eradicate extreme poverty everywhere by 2030.[14] The main current pathway of achieving that, following the rapid industrialisation path of China, would likely lead to much greater inequality within countries, as well as to severe local and global environmental impacts. How to realise access to modern energy services in a way consistent with meeting global climate change mitigation targets will require new thinking and higher levels of global cooperation.

Declining net energy returns

As if the challenges of mitigating climate change, securing economic development and addressing global inequality were not enough, some scientists have argued that we face a further challenge – that of declining net energy returns. As we shall argue in the next chapter, a key perspective followed in this book is that human economic activity is fundamentally dependent on the flows of energy and materials from and to natural ecosystems. The problem of climate change relates to the ability of the biosphere to absorb the wastes produced by human activity, in the form of carbon dioxide and other greenhouse gases. However, the ability of the biosphere to provide useful energy necessary to maintain and increase economic activity presents a further challenge. American systems ecologist Charles Hall and colleagues have argued over the last 30 years that the net energy returns from fossil fuels are declining, and that this could present a fundamental constraint on economic growth.[15] They also question whether renewable energy sources have the capacity to make up for this decline in net energy returns.

This perspective is more controversial and less widely shared than the near-universal acceptance amongst the scientific community of the need to mitigate human-induced climate change, but we think that it is important to consider. We shall look at the challenge of declining net energy in Chapter 10, in relation to insights from other historical and theoretical perspectives for a low carbon energy transformation.

Next steps

The challenge of reducing carbon emissions from energy supply to mitigate human-induced climate change, whilst maintaining the contribution of energy use to economic prosperity and human welfare, then requires nothing less than a complete transformation of our systems of energy supply and use. This is comparable to the scale of past energy transitions, associated with the coal-power industrial revolution and the rise of electricity and oil in the nineteenth and twentieth centuries. So, it should be enlightening to look back at those past energy transitions, in the light of new knowledge and understanding about what drove those transitions, and their economic and social impacts. In the next chapter, we set out the framework that we will use to examine these energy system transitions.

Notes

1 Technically, exergy is defined as the maximum amount of work that can be produced by a system or a flow of matter or energy as it comes into equilibrium with a reference environment. For most purposes, we can take the reference environment as the standard physical state of the environment on Earth (usually taken to be 20°C and 1 atmosphere of pressure). See Carnahan et al. (1975).
2 Schrödinger (1944/2012).
3 Named after James Watt and James Prescott Joule, respectively.
4 IEA (2016b).
5 Arrhenius (1896, 1908).
6 United Nations (2015a).
7 See FRED Blog (2016).
8 ILO (2015).
9 Oxfam (2017).
10 Piketty (2014).
11 Bosworth et al. (2016)
12 Wilkinson and Pickett (2010).
13 Ferrie (2004).
14 United Nations (2015b).
15 Hall and Klitgaard (2012).

3

ASSESSING THE ROLE OF ENERGY IN LONG-TERM INDUSTRIAL CHANGE

Introduction

In order to understand how human social and economic systems have come to be reliant on high quality energy sources, we need useful ways of thinking about these issues. You might have thought that this would be a central concern of modern economic theory. Early eighteenth-century economists were interested in the role that natural resources, particularly land availability, played in the economic system. However, in the late nineteenth and early twentieth centuries, following successive industrial revolutions and the widespread availability of cheap energy sources and the seemingly unlimited land available in North America and British and European colonies around the world, economists began to focus on labour and man-made capital as the key inputs to the economy. This led to the picture of the economy as a closed system. With a few exceptions,[1] it was not until the 1960s and 1970s that ecologically minded economists began to emphasise again the dependence of the economy on natural systems.

After discussing this dependence, we then move on to frameworks for understanding long-term industrial change. Again, mainstream economics provides little guidance, as, for ease of mathematical formulation, it has largely assumed that economic systems are in equilibrium, and the focus is on marginal or incremental changes to these systems. For example, markets consisting of a large number of sellers and buyers of a product or service are assumed to 'clear', i.e. to come to an equilibrium state, in which the quantity and price of product exchanged is optimal and any change would make at least one seller or buyer worse off. This is a useful model if we have, say, a large number of buyers and sellers of bananas, and want to know what would happen if there was a small change in the supply of bananas or people's preferences for eating them. It is not much help when considering long-term and large-scale changes in technologies available and practices of use, which occur in industrial revolutions. Of course, historians have provided many valuable insights as to what has driven such revolutions, which we will draw on. However, in order to more clearly compare different social and economic system changes, and to try to learn lessons to guide us through future system changes, we need a way of examining these changes that is sufficiently general to be applicable, but is sensitive to the detailed and contingent aspects of particular system changes. In the second part of this chapter, we discuss a framework of this type for understanding long-term industrial systems change. First, though, we need to better understand how the processes of energy conversion, discussed in the previous chapter, apply to economic systems.

Ecological dependence of the economy

As noted, mainstream economic theory does not include energy or other resources as drivers of economic growth. The economy is considered as a closed system, not dependent on the natural environment, with households providing labour for firms that accumulate capital (buildings, machinery) in order to provide goods and services that households consume (Figure 3.1).

Economic growth over the long term was then understood as arising from improvement in the amount or productivity of labour and capital in the economy, with natural resources effectively considered unlimited. As this was found not to account for all of the growth in economic output, an additional term representing technological change was added to the equations by economist Robert Solow, though the origin of this technological change was unexplained within the theory.[2] Furthermore, the environmental impacts of economic activity are treated as 'externalities', as they are not one of the factors affecting the agreed price and quantity in the exchange of goods and services between households and firms. For example, until recently, a firm generating electricity from burning coal would not have to pay for the resulting local air pollution or carbon emissions and so the

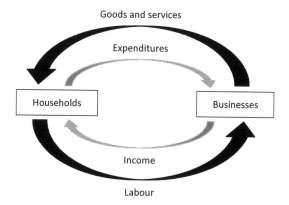

FIGURE 3.1 Circular flow of income in a closed economy.
Source: Adapted from Bureau of Economic Analysis (2014).

value of these impacts would not be reflected in the price that the consumer pays for the electricity. Much of environmental economics has focused on how to 'internalise' these externalities, so that they become part of the economic transaction. This can be done basically in one of two ways: either by regulations that put limits on the level of pollution that a firm can emit, or by putting a price on that pollution through a tax or trading scheme, which affects producer and consumer choices in the transaction. For example, in some countries, including members of the European Union, electricity generators and other industrial plants now have to buy carbon permits under an Emissions Trading Scheme, which is intended to put a value on the carbon emissions, and so give a financial incentive to reduce these emissions. We will discuss the likely effectiveness of this and other incentives to reduce carbon emissions from energy use in Chapter 10. However, some ecologically minded economists have argued that treating the environmental impacts of energy generation and industrial processes in this way fails to fully recognise the dependence of all economic activity on the natural environment.

An ecological economic world view sees the economy as an open system, in which the natural environment provides resources, in the form of primary energy and material inputs, and the capacity to assimilate the wastes produced by economic activity, as well as providing other ecosystem services (Figure 3.2). This view was articulated in a far-sighted essay, 'The Economics of the Coming Spaceship Earth', by British economist Kenneth E. Boulding in 1966.[3] Boulding argued that, though the economy is open in relation to receiving energy from the Sun and in terms of generating knowledge, the economy and the natural environment represents a closed system in relation to the availability and use of other natural resources. He criticised mainstream economists for maintaining a focus on flows of economic activity, whilst assuming that there are 'infinite reservoirs from which material can be obtained and into which effluvia can be deposited'.[4] He called this the mentality of a 'cowboy economy', in which there are unlimited frontiers to be exploited, and

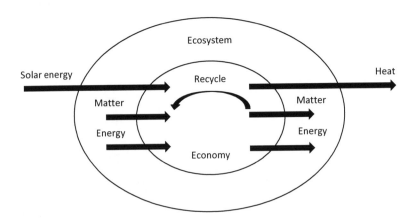

FIGURE 3.2 Economy embedded in natural ecosystem.
Source: Adapted from Costanza et al. (2000).

contrasted it with the need for a 'spaceship economy', in which we should consider how to increase human wellbeing within the finite limits of the planet. This view was reinforced by the first picture of the whole Earth from space, taken by the *Apollo 8* astronauts in December 1968. The later picture of Earth as a 'pale blue dot', taken by the *Voyager* spacecraft in 1990 from 6 billion kilometres away, inspired astronomer and space scientist Carl Sagan's meditation on the precious and fragile nature of the Earth on which we all depend.[5]

More prosaically, economic growth theories then need to consider the contribution of natural resources, particularly energy, to the growth of economic output. The first flowering of this world view in the 1960s and 1970s focused largely on the limits to economic activity that could arise from the depletion of resource inputs, including strategic materials such as copper and oil reserves. This led to the publication of the 'Limits to Growth' study in 1972 which argued, on the basis of computer models, that increasing levels of economic activity due to growth in population and consumption were likely to be constrained by the depletion of resource inputs.[6] Donella Meadows and her fellow authors argued that, unless action was taken to constrain population and consumption growth, this could lead to strains on the global economy with the potential for a collapse in the global economy in the first decades of the twenty-first century. Despite the criticisms from a range of economists, and the acknowledged limitations of the original computer models, recent reviews have found the projections of this study have so far matched well the outcomes in the real world.[7] The limited supplies of rare earth metals and the depletion of fresh water supplies are current key resource input concerns.

However, more recently, ecological economists have focused on the constraints to economic growth that could arise from the limits to the absorptive capacity of the natural environment, and disruption of ecosystem services provided by nature. These include the threat of climate change, due to the increased concentration of greenhouse gases in the atmosphere, but also threats due to the build-up of toxic and disruptive chemicals in the biosphere. The current loss of biodiversity, which is argued to be equivalent to past major extinction events in the Earth's history, threatens to deprive humanity of a range of ecosystem services that nature provides, including new crop strains, pollination of crops and waste dispersion.

This led American ecological economist Herman Daly to propose the concept of a 'steady state economy'.[8] Daly, a former senior economist at the World Bank, became increasingly critical of the focus of mainstream economics on economic growth as the primary means to improving human wellbeing. He argued that, instead, the economy should be designed to maintain a steady flow of materials and energy at levels within the sustainable limits of the Earth to provide these resources and assimilate the wastes produced. Echoing Boulding's ideas, he contended that this reflects that humanity is in danger of overwhelming these limits, as we have gone from an 'empty world' to a 'full world'.[9] We shall return to consideration of these issues throughout the book and, in particular, in Chapter 11.

Systems of production and consumption

In modern mainstream economic theory, the production function plays a key role. This says that the value of economic output for an economy is a function of the inputs to that economy. Classical economists writing in the eighteenth century, such as Adam Smith and David Ricardo, viewed land, labour and capital as key inputs. At that time, the economy consisted primarily of agricultural production and small-scale manufacturing of goods that were traded nationally or internationally. Hence, the availability of land was seen as a key constraint on production. A group of French economists including Marquis de Condorcet and François Quesnay, known as the physiocrats, believed that the wealth ultimately derived from agricultural land, but British economists such as Smith and Ricardo argued that wealth derived from production and market exchange. In his most famous work, *The Wealth of Nations*, published in 1776, Adam Smith introduced the idea of the invisible hand, meaning that desirable collective outcomes can arise out of individuals' decisions without having to be consciously directed. He argued that producers and consumers will act in their own self-interest, but if they are able to freely exchange goods and services in a competitive market, then the outcome will be mutually beneficial. As he put it, 'It is not from the benevolence of the butcher, the brewer or the baker that we expect our dinner, but from their regard to their own self interest.'[10] Philosophers and economists have argued ever since as to the limits and conditions under which this principle holds. However, when there are externalities present, such as unpriced emissions or highly inequitable initial distributions of wealth, then the invisible hand cannot be relied upon to deliver a socially optimal outcome.

Karl Marx, in his three-volume work *Capital*, produced a deep critique of the idea that production and free market exchange would benefit all.[11] He argued that the value of a product should reflect the 'quantity of socially necessary labour' put into its production, but that capitalists typically exploit labour by failing to pay it to reflect the value that it creates. This leads to a process of capital accumulation, in which the owners of capital make profits, some of which are reinvested in further capital, enabling further profits to be made. However, this labour theory of value neglects the value of other inputs, such as land, energy and capital, into production processes. The third classical input into production is capital, which may be defined as any human-made goods that are used in the production of other goods, including machinery, tools and buildings. This reflects the fact that part of economic output is used to enable further production, rather than going to final consumption. This comes from savings which are channelled into investment, rather than being spent on consumption.

In modern neo-classical economic theory, which originated with the work of William Stanley Jevons, Carl Menger and Léon Walras in the 1870s, the production function is retained, but value is defined purely in terms of exchange value, rather than use or production value.[12] In other words, the value of a good or service is only determined by what consumers are willing to pay for it in a freely competitive market. Gradually, as the focus moved to large-scale industrial production and away from agricultural and small-scale manufacturing production, 'land' was largely dropped as a factor of production, leaving only 'labour' and 'capital'.

As we can see, energy plays no role as a factor of production in any of these perspectives. Partly, this reflects a historical accident. The concept of energy had not been formalised when the classical economists were writing in the eighteenth century, and fossil fuels had not yet begun to be used in any quantity. It was only just becoming clear when the early neo-classical economists were writing in the second half of the nineteenth century, and they ignored any constraints from energy or other resources in their theoretical formulations.

The Entropy Law and the Economic Process

The first modern comprehensive view of the energy and ecological dependence of economic activity was developed by Romanian economist Nicholas Georgescu-Roegen in his major work *The Entropy Law and the Economic Process*, published in 1971.[13] As the title suggests, Georgescu-Roegen applied the second law of thermodynamics, the entropy law, to understanding economic processes. He argued that, just as living systems require a flow of low entropy energy to survive and reproduce, economic systems should also be understood as dependent on flows of low entropy energy that is converted to produce goods and services, whilst giving rise to high entropy wastes. He was clear that this is not merely an analogy. Real economic processes, such as the production of steel and chemicals or even the processing of information, require a flow of low entropy inputs in the form of ordered energy and raw materials, such as oil and iron ore. These are converted into useful products, such as steel and chemicals, in industrial processes. In these processes, wastes are created, including high entropy heat, air pollutants, such as carbon emissions, and material wastes. At a macro level, the whole economy can similarly be understood as a system for converting low entropy inputs into useful goods and services involving creation of high entropy wastes.

Though Georgescu-Roegen's work was one of the foundations for the approach of ecological economics, it was largely ignored by mainstream economics. This can be seen as resulting from historical, ideological and theoretical reasons. Despite the oil price increase shock in 1973, shortly after the publication of his book, mainstream economic thinking at that time was (and still is) dominated by neo-classical ideas based on the fundamental notion of exchange value rather than production or use value. Even in mainstream production and growth theory, the production function was largely thought of as an accounting tool, rather than representing biophysical conversion processes. These ideas underlay the rise of free market economic thinking, inspired by the theories of Friedrich Hayek and Milton Friedman, that came to dominate political and economic debate in the 1980s and 1990s, particularly after oil prices returned to lower levels.

Georgescu-Roegen also contributed to the sidelining of his ideas by committing, in his later work, what even most ecological economists now think was a theoretical mistake. He emphasised the degradation of both energy and materials in economic processes. This led him to speculate what he called a fourth law of thermodynamics – that complete recycling of matter is impossible. However, ecological economists, such as Robert Ayres, pointed out that there is a key difference between energy and matter in physical terms.[14] The second law of thermodynamics does imply that complete recycling of energy is impossible. The ordered, concentrated energy in coal and oil is degraded to high entropy heat when it is used to provide useful work, e.g. to power an engine. This high entropy heat cannot be recycled without a further input of low entropy energy, for example, to create a temperature difference that can then be used to power a heat engine. On the other hand, given enough ordered energy, matter can be recycled indefinitely. The stores of ordered materials on the Earth, such as iron ore, are the result of natural geological processes. In economic production, these are converted to useful materials such as steel which go into products. Eventually, these products become wastes and are disposed of in landfill, for example, where they are mixed with other product wastes. However, even then, given enough ordered energy, they can be recovered, separated and reused. The limits to the recycling of matter are thus related to the availability of cheap energy and the practicality of recycling. This is important for ideas of a circular economy, in which products are designed to be reused or remanufactured, as we shall discuss later.

None of this should undermine Georgescu-Roegen's key contribution in understanding that economic processes rely on the conversion of low entropy energy to high entropy heat. Of course, economic processes cannot be understood solely in biophysical terms. The aim of economic processes is to produce goods and services that satisfy human needs and wants and so contribute to human wellbeing. For a full understanding of economic activity, we need to understand how the consumption of goods and services contributes to improving human wellbeing. We shall return to this topic in Part IV of the book. However, in the following parts of the book, we aim to show how biophysical drivers add to, and interact with, other social and technological drivers of industrial change.

Energy as an input to economic systems

Georgescu-Roegen's work inspired a new generation of ecological and biophysical economists, including Cutler Cleveland, Robert Costanza, Charles Hall and Robert Kaufmann. In 1984, they published an article in the leading journal *Science* on the energy dependence of economic activity. They noted that 'Production is explicitly a work process during which materials are concentrated, refined, and otherwise transformed. Like any work process, production uses and depends on the availability of free [ordered] energy.'[15] Looking at over 90 years of data for the US economy, they showed that there was a very close correlation between primary energy input to the economy, mainly in the form of fossil fuel use, and the economic output of the economy, measured by GNP. Though correlation is not causation, their findings have been supported by subsequent analyses for the US and the global economy (see Figure 3.3), and it has been argued that the causation

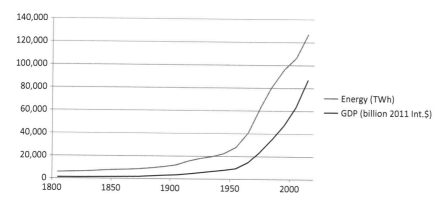

FIGURE 3.3 Global energy use and economic activity, 1800–2008 (GDP in billion 2011 International $).
Sources: Author own figure, based on data from Smil (2010) (energy use) and Maddison (2010)/World Bank/Our World in Data.

works both ways, i.e. that more energy use is a driver of economic activity, and that higher economic output stimulates the further extraction and use of energy.[16] In particular, Cleveland and colleagues argued that this helps to explain a key fact in economic thinking: increases in labour productivity, i.e. the economic output per person or per hour worked, are understood to be a key driver of economic growth. One enabler of labour productivity improvement is that labour is able to produce more output thanks to the use of machines powered by energy. Thus, as machines, and so energy, substitute for manual labour, total economic output rises. Provided that the additional economic activity stimulated in this way gives rise to additional job opportunities, total employment can increase.

Cleveland and colleagues pointed out that if the role of energy in substituting machines for labour is significant, as this evidence suggests, then the availability of ordered energy as an input to the economy is crucial. Furthermore, they pointed out that it is not the aggregate amount of energy available that is important, it is the energy surplus, after taking into account the energy used in the process of extraction of energy sources. They showed that this energy surplus has been declining, since peaks for oil and coal in the 1940s and 1950s of over 80 units of energy returned for each unit of energy invested in extraction. This energy surplus was down to 23 for production of oil and gas in the US in the 1970s, and initial values of the energy surplus for renewable energy sources were even lower.[17] This is because, despite the high flows of solar energy reaching the Earth, this solar energy is much less concentrated than the energy in fossil fuel sources, and so requires more energy in extraction per unit of energy output. The energy surplus values for renewables, though, can increase with technological improvements that enhance the efficiency of conversion of natural energy flows to useful energy.

Charles Hall and colleagues subsequently argued that if energy surplus of fossil and renewable sources does continue to decline into the future, then more and more economic output will need to be invested in the extraction and processing of energy. This would mean that less is available for investment and consumption of discretionary goods, as opposed to necessities, which would mean the slowing of economic growth.[18]

We shall return in Chapter 10 to the interaction between economic activity, energy surplus and constraints on carbon emissions to address climate change. We now turn to approaches to understanding industrial change, so that we can better understand how the availability and use of energy interacted with other drivers of change starting from the industrial revolution.

Creative destruction

Throughout most of human history, the scale of economic output grew very little. Then, starting with the industrial revolution in Britain in the eighteenth century, economic growth took off. Between 1900 and 2008, global GDP per person increased by a factor of six from $1,260 to $7,600, whilst world population also increased from 1.5 billion people to 6.8 billion people, thanks to technological and social advances.[19] This means the size of the global economy increased by an astonishing 25 times from under $2 trillion in 1900 to over $51 trillion (in 1990 Int.$) in 2008. Various theories and ideas have been proposed to explain this unprecedented rate of economic growth, though none as yet provides a complete explanation, and different ideas may be mutually reinforcing.

Classical economists, such as Adam Smith and David Ricardo, emphasised the role of division of labour and gains from trade. The notion of comparative advantage holds that if nations specialise in producing what they are most efficient at producing, and trade the surplus with other countries that are able to produce different specialities more efficiently, then all parties will be better off. As the size of markets grow, this enables further efficiency improvements from economies of scale in production processes, creating a positive feedback loop driving economic growth.

In the first half of the twentieth century, the Austrian-born economist Joseph Schumpeter made a revolutionary contribution by emphasising the role of innovation in economic growth. In his early work, *The Theory of Economic Development*, published in German in 1911 and in English in 1934, he emphasised the role of entrepreneurs who develop and introduce new technologies and new ways of organising production.[20] As these diffuse through the economy, they create value and thus drive economic growth. In his later work, on *Capitalism, Socialism and Democracy*, published in 1942, he emphasised the role of large firms conducting dedicated research and development (R&D) to produce innovation.[21] Most famously, he used the metaphor of 'creative destruction' to describe structural change in economies, in which old firms and technologies are replaced by new firms and technologies that offer new values and new opportunities to consumers.

He thus rejected the role of equilibrium which was central to mainstream economic thought. These ideas have led to a rich seam of work examining the processes of technological, industrial and economic change, sometimes called neo-Schumpeterian or evolutionary economics.

The role of institutions

Another important line of economic thinking focuses on the role of institutions in driving or constraining economic change. In an influential recent book, US economists Daron Acemoglu and James Robinson argue for the role of political and economic institutions in determining why some countries got rich and others stayed poor.[22] They contend that this requires inclusive political institutions, which support political freedoms and distribute political power through society but nevertheless have sufficiently centralised states to maintain law and order and protect these freedoms, and that these reinforce inclusive economic institutions. These inclusive economic institutions are the economic rules that promote innovation and entrepreneurship, by providing confidence to entrepreneurs that they will be able to benefit from their ideas and these will not be appropriated by the state or other powerful individuals. These institutions include secure private property, independent legal systems and the provision of public infrastructure and welfare safety nets. Acemoglu and Robinson argue that countries and regions such as the US, Western Europe and Japan (after the Second World War) that have inclusive institutions are the ones that experienced high rates of economic growth, so their citizens became richer, whereas developing countries that had more exclusive institutions (often partly as a result of the disastrous effects of European colonisation) stayed poor. However, they recognise that this historical development was a path-dependent process, in which decisions made at critical junctures often had significant long-term impacts. For example, they cite the Glorious Revolution in England in 1688 which secured parliamentary authority over economic institutions, led to the adoption of a Bill of Rights in 1689, and spurred the development of incentives for investment, trade and innovation. Acemoglu and Robinson built on the work of other economists and political scientists, such as Douglass North and Paul Pierson, who emphasised the role of institutions, such as private property regimes and the granting of limited liability to companies, in stimulating economic growth and path-dependent structural change.[23] However, this type of institutional analysis also did not recognise any role for energy or ecological dependence in explaining economic growth.

Technological revolutions

In this book, we seek to combine the ecological economics perspective set out above with an evolutionary and institutional economics understanding of dynamics and change within economies. A pioneer in this approach was Chris Freeman, who was the founder and first director of the Science Policy Research Unit (SPRU) at the University of Sussex, UK, where I now work. Building on the work of Schumpeter, Freeman was interested in the economics of industrial innovation, and how this leads to change at the macro level of whole economies.[24] At this macro level, he saw economies evolving through interactions, or 'coevolution', between five domains: scientific knowledge, technological developments, economic systems, political systems and culture. He articulated this coevolutionary history of economic change, consisting of five 'long waves' dating back to the industrial revolution, in a 2001 book *As Time Goes By* with Francisco Louçã.[25] We draw on the fascinating historical insights in that book in the present work.

Working both independently and in collaboration with Chris Freeman, a new theoretical model of technological revolutions and their economic implications was developed by Venezuelan economist Carlota Perez.[26] This model and its implications were set out in detail by Perez in her 2002 book on *Technological Revolutions and Financial Capital: The Dynamics and Bubbles of Golden Ages.*[27] She argued that global economic growth has gone through five distinct surges, each associated with a technological revolution, but only fully realised when institutions and practices adapt to take advantage of the new opportunities offered. Her model emphasises the interconnectedness of technological systems and the associated network of researchers, engineers, suppliers, producers, users and institutions.

Perez argues that each of these five surges of development was initiated by a technological breakthrough, beginning with Arkwright's first mechanised mill for spinning cotton in 1771, which helped to kick-start the industrial revolution (see Table 3.1). However, the first installation phase of the surge is dominated by financial capital, leading to a frenzy of speculative investment, as was the case with the building of canals to transport cotton and other goods during the 'canal mania' in the 1790s. Only after this financial bubble bursts could a more stable phase of deployment occur, led by productive capital, resulting in a 'golden age' of high economic growth. Crucially, in each surge, the new technologies and organising principles not only created new industries, but also changed the 'common-sense' shared ideas of what constitutes profitable innovation and investment. Perez calls this new common-sense set of a shared ideas a 'techno-economic paradigm'. The adoption of a new paradigm, following a structural crisis of adjustment, involves: (1) changes in relative cost structures, as new low-cost technologies and infrastructures create cost advantages for investment; (2) changes in opportunity spaces, as entrepreneurs seek to take advantage of these new opportunities; and (3) new organisational and business models that become accepted best practice.[28] These factors mutually reinforce each other to create a favourable direction for innovation and technological change. This is strongly contested in the installation phase by firms and investors who were still ben-efiting from assets from the previous surge. However, as the benefits of the new paradigm become more widely shared in the deployment phase, this paradigm becomes much more widely accepted. This results in a golden age of high economic growth, until the paradigm begins to hit diminishing returns, and a new surge arises.

We find Perez's model very enlightening and we broadly follow it in our dis-cussion of surges of industrial change in subsequent chapters. However, there are

TABLE 3.1 Perez's five successive technological revolutions, 1770s to 2000s

Technological revolution	Name	Lead country or countries	Initial event	Year
First	The 'Industrial Revolution'	Britain	Arkwright's mill opens	1771
Second	Age of Steam and Railways	Britain	Test of the *Rocket* railway steam engine	1829
Third	Age of Steel, Electricity and Heavy Engineering	US and Germany	Carnegie Bessemer steel plant opens	1875
Fourth	Age of Oil, the Automobile and Mass Production	US	Ford Model-T car plant opens	1908
Fifth	Age of Information and Communications Technologies	US	Intel microprocessor chip produced	1971

Source: Based on Perez (2002), p. 11, Table 2.1.

two caveats to this model. First, it places too much emphasis on the role of technologies in driving industrial and economic change. Perez herself emphasises that this should not lead to a charge of 'technological determinism', i.e. that history is seen as the unambiguous advance of technological progress. From an evolutionary economics perspective, the adoption of particular technologies and industries depends on social choices and how these are embedded in institutional rule systems. As historians are keen to point out, different choices could have been made and technological and institutional change could have gone in different directions. Indeed, some have argued that a radically different model of industrial development did flourish for a while in some countries, before being overwhelmed by the capitalist industrial model.[29] So, it is important to focus on these choices and critical junctures when we examine the past waves of industrial change.

Second, though it does recognise the significant role of energy inputs which become low cost in different surges, the model does not connect with the ideas on the energy dependence of economic growth. One of the reasons for writing this book is to try to build a bridge between these two sets of ideas on industrial-based economic change and ecological dependence of economic activity.

Coevolutionary approaches

In order to bring together these two sets of ideas, we develop a model that is based on the interaction or coevolution of different subsystems. Evolutionary economics developed in the 1970s and 1980s to take forward the insights of Schumpeter on innovation and economic change. In particular, in their 1982 book *An Evolutionary Theory of Economic Change*, Richard Nelson and Sidney Winter argue against the mainstream economic view of firms and individuals as fully rational with perfect foresight, who are able to make optimal decisions on what they produce and consume, according to their preferences.[30] Instead, Nelson and Winter start from the more feasible assumption that firms and individuals have 'bounded' rationality, i.e. they are not able to collect or process all the information relevant to any decision, and so they have to search for routines that generate satisfactory outcomes. These routines are then selected by internal firm processes and external selection processes, which are influenced by social and institutional factors, as well as perceptions of likely profits. Outcomes are then not optimal, as a naive view of evolutionary change might suggest, but they depend on the bounded rationality of the firms and individuals involved and on how they interact with other technological, natural and institutional systems. These interactions give rise to constantly changing selection environments. In this way, change is path dependent with outcomes depending on the results of past choices and how they were embedded in institutions.

This leads to the idea that technologies and systems can become 'locked in' and resistant to change. This was first articulated by economist Brian Arthur and economic historian Paul David. Arthur argued that increasing returns or positive feedbacks to adoption mean that the more people adopt a technology, the more likely it is to be further adopted.[31] One driver of this is network economies – as more people in your social network have a particular technology, such as a mobile phone, then the more benefit you get from adopting that technology. Drawing on complex systems ideas, he argued that small random effects in the early stages of adoption of a technology could be amplified by these positive feedback effects, leading to lock-in of the winning technology, whilst an alternative gets shut out. Working independently, David studied historical examples of this, most famously the QWERTY keyboard layout.[32] He showed that as different typewriter manufacturers copied this layout to meet the needs of typists already trained on this layout, it quickly became the industry standard. Despite the claim that other keyboard layouts were more ergonomically efficient, the QWERTY layout remained locked-in on computers and now even on the virtual keypad on your mobile phone.

Moving from particular technologies to the level of technological systems, Gregory Unruh noted that technological choices interact with the institutional rule systems that structure human interactions. High carbon fossil-fuel based energy technologies, such as coal-based electricity generation and oil-based transportation, have benefited from positive feedbacks to their adoption, including economies of scale and learning and network effects. Moreover, the rule systems governing their use have benefited from similar positive feedbacks. For example, the rules governing electricity networks have been written based on firms using large centralised generation plants and, as firms adapt to these rules, they would not want to change to another set of rules. Alternative systems, such as more distributed energy systems based on renewable technologies, then face high barriers to their adoption. Hence, carbon-based energy systems have benefited from this interaction or coevolution between the technologies and the institutional rules, leading to what Unruh called 'carbon lock-in'.[33]

Work on the coevolution of technologies and institutions was also undertaken by the evolutionary economist Richard Nelson. He applied this perspective to the development of industrial systems, and argued that it forms the underlying driver of economic growth at the macro level.[34] This work was taken forward by Johann Peter Murmann at the industry level and Eric Beinhocker at the macro level. Murmann examined the development of the chemical dye industry in Britain and Germany at the end of the nineteenth and early twentieth centuries in terms of the coevolution of technologies, institutions and firms' strategies.[35] He found that although many of the key technologies were developed in Britain they were successfully commercialised by German firms, partly because of the close links between the universities and the newly developed industrial research laboratories in Germany. This reinforced incentives for firms to follow business strategies based on continuing innovation.

In his book *The Origin of Wealth: Evolution, Complexity and the Radical Remaking of Economics*, Beinhocker brought together the ideas of Nelson and Arthur to propose that economies can be seen as complex adaptive systems.[36] He argued that modern economies have evolved through a process of coevolution of technologies, institutions and business strategies, and it is the mutual positive feedbacks to the adoption of these that have driven economic growth. For example, Arkwright's invention of the technology of the spinning frame in the 1770s stimulated the institutional innovation of organising the manufacturing of cloth in large factories, rather than cottage industries, and the business strategy associated with the employment of labour and machinery in the factory, so as to make a profit. The wealth enjoyed by citizens of Western countries then reflects the enormous expansion in the provision and variety of goods and services that a modern economy can supply, as a result of these coevolutionary processes driving growth. We shall return to a discussion of how we can understand the processes driving economic growth once we have examined in more detail the major historical waves of industrial change through which these coevolutionary processes played out.

Beinhocker aims to take into account the ecological economics ideas of Georgescu-Roegen, but we argue that he misinterprets these ideas in his discussion of the role of entropy in production processes. Beinhocker argues that production involves a flow of 'high entropy to low entropy', interpreting production as a process of order creation. This is correct, to a degree, in that the results of the production process are indeed more ordered, such as the steel produced from iron ore with energy inputs. However, it ignores the vital role that flows of energy and resources play in production. Georgescu-Roegen argued that local entropy reduction in production processes needs to be 'financed' by a flow of low entropy energy from natural systems, which leads to the creation of high entropy wastes.[37] We argue that this strengthens the case for combining an ecological perspective with an evolutionary understanding of the processes of economic change.

As a final step to be able to do this, we need a framework that brings together these coevolutionary dynamics of industrial change with the energy and ecological dependence of economies. Drawing on the above ideas on coevolutionary dynamics and related ideas within ecological economics,[38] a simple framework of this type was developed by the present author.[39] This brings in two further

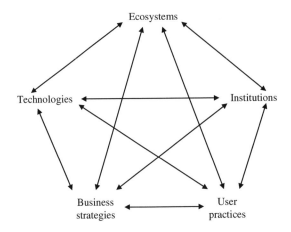

FIGURE 3.4 Coevolutionary framework.
Source: Foxon (2011, p. 2262).

elements to the above coevolutionary approach (Figure 3.4). First, the framework includes the system of user practices, to recognise that end users of a technology play a significant role in the evolution of that technology. For example, the design of new electrical devices such as washing machines has been heavily influenced by how practices for washing clothes have changed to take advantage of the new opportunities offered. Second, the framework includes ecological systems, to reflect particularly the influence of the availability of energy sources and the ability of the natural environment to absorb wastes on the development of industrial systems. Thus, in this framework, economic change is seen as a process of coevolution of technologies, institutions, business strategies, user practices and ecosystems.

Other related approaches

Three recent studies cover some of the same ground as this work, but from different perspectives. First, in *The Subterranean Forest*, German social and environmental historian Rolf Pieter Sieferlie examined the role of energy in the different processes of industrialisation in England and Germany in the eighteenth century.[40] In a rich historical study, also drawing on insights from ecological economics, he showed how access to relatively cheap coal, the subterranean forest of the title, together with a nascent capitalist system and supportive political institutions, helped to drive the coal-based industrial revolution in England. In contrast, in Germany, despite a perceived wood fuel crisis, the lack of good transport links for coal, concerns on its impact on health and the bureaucratic and politically divided German states served to delay the transition to the use of coal and wider industrialisation.

Second, in *Heat, Power and Light*, British energy economist Roger Fouquet has collected and analysed a wealth of historical data on the uses and prices of energy services in Britain back to 1450 to chart revolutions in the provision of those energy services of heating, power and lighting.[41] As we shall discuss in more detail in Chapter 8, he argued that, in the long run, technological innovations enabled reductions in the cost of energy services, stimulating increasing demand for those services. Crucially, this view also shows how new energy technologies improved the efficiency of providing those services, significantly improving the quality of the service to household and business users. For example, he shows that it was the reduction in the cost of coal for providing heating and power, together with the higher quality service it provided, that drove the substitution of coal for wood from the early 1600s.

Third, in *Power to the People*, Swedish, Italian and British economic historians Astrid Kander, Paolo Malanima and Paul Warde examine the role of energy in the economic history of Europe.[42] They identify three industrial revolutions relating to the use of energy. Their first industrial revolution, starting from around 1770 in the UK with the expansion of first water and then steam power for industrialisation, corresponds to Perez's first and second surges. Their second industrial revolution, starting from around 1870, relating to the rise of electricity and oil, corresponds to Perez's third and fourth surges. Their third industrial revolution, starting from 1970, relates to the rise of information and communication technologies (ICT), the transition from predominantly manufacturing to service-based economies in Europe and the consequent partial decoupling of energy use from economic growth. This corresponds to Perez's fifth surge and relates to current challenges of decarbonising economies that are covered in this book. They use the concept of 'development blocks' to describe the systems of technologies, infrastructure, energy sources and institutions that drove these revolutions, leading to phases of high economic growth. Their work is thus very much complementary to the coevolutionary approach that we have used in this book, and they provide fascinating insights by putting together reliable numbers on energy consumption over these periods.

The reader seeking more detailed historical analysis and quantitative data on the relations between energy use and economic growth is urged to refer to these works, though hopefully they will still find some novelty in the current book.

Fourth industrial revolution

The term Fourth Industrial Revolution has also been used by the founder and chair of the World Economic Forum, Klaus Schwab, and so has been widely taken up in business and political debates.[43] Schwab uses this phrase to contrast with the three previous industrial revolutions, relating to the use of water and steam power, electrification and information and communication technologies to advance production processes, as in Kander and colleagues' classification. He argues that the Fourth Industrial Revolution, involving interactions between technological developments in artificial intelligence, robotics, 3-D printing, nanotechnology, biotechnology, materials science and energy storage, is evolving at an exponential pace and has the potential to disrupt every current industry. He recognises that society has the power to shape how these technologies are developed and used, not just to respond to them, and that it needs to take into account the impacts on those such as workers in declining industries who may be left behind. We welcome this debate, but feel that it could be informed by a better understanding of the drivers of previous industrial revolutions and the role of energy sources and conversion in those revolutions, which this book aims to contribute to.

With these theoretical building blocks in place, we can go on, in Part II, to examine the role of energy sources and technologies in past long waves of industrial change, leading to surges of economic growth, by applying the coevolutionary framework.

Notes

1　In 1926, British Nobel-prize winning chemist Frederick Soddy published a book *Wealth, Virtual Wealth and Debt* (Soddy, 1926), in which he argued that real wealth is derived from the use of energy to transform materials into physical goods and services. This foreshadowed the later work of Nicholas Georgescu-Roegen, but made little impact at the time.
2　Solow (1956, 1957).
3　Boulding (1966).
4　Boulding (1966), p. 4.
5　Sagan (1994).
6　Meadows et al. (1972).
7　See Turner (2008, 2014).
8　Daly (1977/1991).
9　Daly (2015).
10　From *The Wealth of Nations* (Smith, 1776/1999), Book 1, Chapter 2, p. 119.
11　Marx (1867–1894).
12　The modern mathematical basis for neo-classical economics was skilfully formalised by American economist Paul Samuelson in his 1946 book *Foundations of Economics* (Samuelson, 1946) and communicated to students in a textbook, first published in 1948 and now in its 19th edition (Samuelson, 1948; Samuelson and Nordhaus, 2009).
13　Georgescu-Roegen (1971).
14　Ayres (1997).
15　Cleveland et al. (1984), p. 890.
16　Csereklyei et al. (2016).
17　Energy surplus or net energy is usually measured by the 'energy return on energy invested', or EROI, see Cleveland et al. (1984); Hall and Klitgaard (2012).
18　Charles Hall and colleagues developed a simple model, called the 'cheese slicer model' to illustrate this, see Hall et al. (2008), Hall and Klitgaard (2012).
19　Maddison (2010).
20　Schumpeter (1911/1934).
21　Schumpeter (1942).
22　Acemoglu and Robinson (2012).
23　North (1990); Pierson (2000).
24　Freeman and Soete (1997); Freeman (2008).
25　Freeman and Louçã (2001).
26　Perez (1983); Freeman and Perez (1988).
27　Perez (2002).
28　Freeman and Perez (1988).
29　Schot and van Lente (2010).

30 Nelson and Winter (1982).
31 Arthur (1989).
32 David (1985).
33 Unruh (2000).
34 Nelson (2005, 2008).
35 Murmann (2003).
36 Beinhocker (2006).
37 As Georgescu-Roegen puts it, 'Since the economic process materially consists of a transformation of low entropy into high entropy, i.e. into waste, and since this transformation is irrevocable, natural resources must necessarily represent one part of the notion of economic value' (1972, p. 28).
38 Kallis and Norgaard (2010); Norgaard (1994); Maréchal (2007).
39 Foxon (2011).
40 Sierferlie (1982/2001).
41 Fouquet (2008).
42 Kander et al. (2014).
43 Schwab (2016).

PART II

Long waves of energy-industrial change

4

ENERGY USE IN PRE-INDUSTRIAL SOCIETIES

Introduction

In this chapter, we start by briefly reviewing the technological and social innovations relating to the development of agricultural societies. This required greater control of flows of solar energy collected by plants, and enabled much higher population densities, providing the basis for future industrial developments.

From hunter-gatherers to agriculturalists

For most of human history, people lived a hunter-gatherer lifestyle. The majority of their food energy came from gathering a range of natural plant sources. This was supplemented by the hunting of large and small prey animals. This lifestyle was often nomadic, so that when the available food sources in a given area were depleted, they could move on to another area. Though there was always the threat of violent conflict with neighbouring tribes, it has been suggested that groups would also act to limit excessive population increases, so that they maintained a buffer in case of hard times.[1] This meant that when food was plentiful hunter-gatherers could obtain all the calories they needed without excessive effort. They could then spend the remainder of their time in social activities. The main additional energy source used by hunter-gatherers was the burning of wood to produce fire. This was primarily used for cooking meat and vegetables, as well as providing warmth in cooler climates. There is evidence that the use of cooking improved the quality and range of food available, and so may have been an important contributor to improvements in brain capacity and social skills.[2]

Around 12,000 years ago, a significant change to this lifestyle began to occur, which is referred to as the Neolithic Revolution. This is the adoption of agriculture, which occurred almost simultaneously in the Middle East, East Asia and

Central America. This involved the deliberate cultivation of plants and the husbandry of animals, which included selective breeding to enhance characteristics favourable to humans, such as the relative size of edible grains produced by grasses such as wheat and rice, and the docility of pack animals such as sheep and cows. This then required groups to remain fixed in one location. The main advantage of this new agrarian lifestyle was that it permitted higher population densities, so more people could live in a given area. This was because of the higher quality and quantity of food collected, due to the relatively higher calorie count of selected plants and animals, and the ease and security of these domesticated food sources. The main disadvantage was that these higher populations meant that more time and effort was required to be exerted to tend and harvest these food sources. So, although more food was available, most people had to work longer and harder to produce it, compared to their hunter-gatherer forebears. Contrary to ideas of universal progress, the wellbeing of most people was thus probably lower after the Neolithic Revolution.

Coevolutionary perspective

In terms of ecological dependence within our coevolutionary framework, agricultural societies were still dependent on flows of solar energy collected by plant biomass, and used for food and fire for cooking and warmth. The main difference from hunter-gatherer societies was the enhanced efficiency of harnessing this energy. These societies have thus been characterised as 'controlled solar energy systems'.[3] This was enabled by and stimulated further technological and social changes. The natural within-group sociability of humans enabled people to live in close proximity with the high population densities referred to. The technological improvements in terms of the managing of natural processes, such as selective breeding, relied on existing knowledge of plant and animal characteristics and behaviours, built up over millennia. Through a learning process of trial and error, this knowledge was developed and passed on to future generations through cultural learning, which is a much faster process than natural biological selection. This also enabled social changes in terms of institutional rules and user practices (though not yet business strategies), as we shall discuss below.

By deliberately altering the characteristics of plants, animals and ecosystems in this way, e.g. by clearing woodland to create field and grasslands, humans were able to significantly enhance the proportion of natural biomass that could be used for feeding people and livestock.[4] This can be measured in relation to the net primary production or NPP of natural ecosystems. It has been estimated that whereas hunter-gatherer societies consumed 1 per cent or less of the available NPP of the ecosystems they inhabited, agricultural societies could consume over 75 per cent of the available NPP of their modified ecosystems.[5] This enabled a much higher level of energy and material use per person (see Table 4.1).

This view of the ecological dependence of human societies is summarised in the term 'social metabolism'.[6] This represents the entire flow of materials and energy

TABLE 4.1 Metabolic profiles of hunter-gatherers and agricultural and industrial societies

	Unit	Hunter-gatherers	Agricultural society	Industrial society
Total energy use per person	GJ/yr	10–20	40–70	150–400
Population density	/km^2	0.025–0115	up to 40	up to 400
Agricultural population	%	–	more than 80%	less than 10%
Total energy use per unit area	GJ/ha/yr	up to 0.01	up to 30	up to 600
Biomass (share of energy use)	%	more than 90%	more than 95%	10–30%

Source: Adapted from Haberl et al. (2011).

that are required to sustain all human economic activities. As noted in Chapter 3, all complex biophysical processes rely on a flow of ordered energy to maintain their structures and activities. The term 'social metabolism' is thus introduced by analogy to the biological metabolism of the processes in animals and humans by which a flow of ordered energy is used to maintain bodily functions and to reproduce. However, social metabolism goes beyond this by linking material and energy flows to social organisation. Hence, different socioeconomic systems vary in the quantity and quality of resource use and the sources and sinks of the output flows.[7] In addition to the feeding of people and livestock, the social metabolism of agricultural systems includes the energy and raw materials needed for tools and equipment such as ploughs, and buildings and other infrastructure. In turn, the higher energy and material use of agricultural societies stimulated increased social innovation to manage the higher complexity of these societies.

Social innovation in agricultural societies

The higher yields and higher population densities of agricultural societies created challenges in terms of the management and distribution of the food sources, which drove innovations in the organisation of societies. Excess harvests could be stored to smooth out variations in grain production from year to year, but this required protection of these stores and also questions of how this excess was to be distributed between members of the society. This contributed to the development of two key social innovations: hierarchical organisation and division of labour.

In order to both manage internal distribution processes and protect against other neighbouring groups who might be tempted to steal the stores, agricultural societies needed more complex organisational structures. These were typically hierarchical structures, based on strong leaders who maintained order and control and protected the society from aggressors. These evolved from family clans but required unrelated individuals to work together for the common good. However, leaders were able to accumulate power and so sometimes to act in the self-interest of themselves and their supporters.

The second innovation resulted from the fact that more complex organisational structures created the potential for members of the community to specialise in particular activities, such as tool making, and to trade what they produced for access to the community's food sources and other resources. This began the process of division of labour that Adam Smith described as the source of complex economic interactions.

Tragedy of the commons

As suggested previously, the Neolithic Revolution enhanced the potential for conflict between individual gain and the wider social or common good. Though this has been discussed by philosophers throughout the ages, in relation to conflict over scarce land and resources this was highlighted in evocative form in 1968 by Garrett Hardin, who coined the phrase 'tragedy of the commons'.[8] This described the potential for overgrazing of common land, as each individual has the incentive to allow their animals to graze, but the collective outcome of this could be harmful to all. In fact, work by Elinor Ostrom and colleagues showed that many traditional societies had evolved institutional rules governing grazing of common land to prevent this negative outcome, with community punishments for transgressors.[9] As societies became more complex, these rules and systems of punishments needed to evolve to prevent 'free-riding' by individuals against the common good.

Sustainability of agricultural societies

As noted in Chapter 3, any society needs to generate an energy surplus, i.e. a positive net energy return, in terms of food, firewood and other useful energy sources such as animal dung, which must be greater than the energy expended to produce this, in the form of human and animal labour in this case. For agricultural systems, to take into account losses and inefficiencies in energy conversion processes, this energy surplus needs to be at least 5:1, i.e. the energy return needs to be at least five times the energy expended.[10] Because there are limits to the potential for improvements in productivity of these natural-based processes, this creates limits to the potential for further increases in population density of agricultural societies. As we shall see later, the use of fossil-fuel based fertilisers enabled significant agricultural productivity improvements in the 1960s and 1970s, but at the cost of adding to local and global environmental impacts.

Ongoing agrarian–industrial transformation

In our broad sweep of history, the next crucial stage is the beginning of the industrial revolution in Britain in the eighteenth century, which led to the widespread use of fossil fuel energy sources, as we shall describe in forthcoming chapters. However, it is important to remember that the benefits of this industrial revolution have not yet spread globally. In many so-called developing countries, the majority of their populations are still engaged in subsistence agriculture and rely mainly on biomass energy sources to meet their energy needs. This has important consequences for the potential for a future sustainability transition, as we shall discuss.

However, we now move on to eighteenth-century Britain.

Notes

1 Sieferle (1982/2001).
2 Wrangham (2009).
3 Sieferle (1982/2001).
4 Haberl et al. (2011).
5 Boyden (1992).
6 The development of thinking on society's metabolism and material flow analysis is discussed by Fischer-Kowalski (1998) and Fischer-Kowalski and Hüttler (1998).
7 Haberl et al. (2016).
8 Hardin (1968).
9 Ostrom (1990).
10 Hall et al. (2009).

5

THE FIRST INDUSTRIAL REVOLUTION – WATER AND STEAM POWER

Introduction

This chapter examines the historical surges associated with industrialisation, and the use of water and steam power from coal. Positive feedbacks from the application of knowledge, economic incentives for investment and trade, and the availability of natural resources helped to drive these surges in Britain.

Why in Britain?

Most historians agree that what happened in Britain between around 1750 and 1850 was significant enough to be called the 'Industrial Revolution', involving new mechanically powered industrial production, a new source of power in steam power from burning coal and the application of steam power to transport in the form of railways. However, historians and economists disagree on the main drivers of these changes and on the timing and significance of the economic effects of these innovations. In particular, why did it happen in Britain and at that time? After all, in the preceding centuries, China could justifiably claim to have been more scientifically advanced, and other European countries, such as the Netherlands, had higher economic output per person before this time. We cannot come to definitive answers to these questions here, so we focus on our core topic of the interactions between new sources of energy and other social and institutional changes. In terms of timing, we broadly follow Perez's notion of two surges, with the 'installation phase' of the first surge beginning in 1771 with the opening of the first (water-powered) mechanical mill for spinning cotton, and the 'installation phase' of the second surge beginning with the testing of the *Rocket* steam engine for the Liverpool–Manchester railway in 1829. However, it is important to begin by tracing the economic and cultural context in Britain that enabled these innovations to be developed and to take root.

Three key precursors have been identified that meant that Britain was economically and intellectually receptive to new ideas and their implementation. First, the scientific revolution started by Galileo and Newton created the conditions in Britain for the application of new knowledge to practical challenges (though the practical understanding often ran ahead of the detailed scientific explanations). Second, political and economic changes contributed to the development of Britain as a trading nation with stable and inclusive political institutions. These created incentives for the utilisation of knowledge to create economic wealth. Third, the natural resources available in Britain in the form of water power, and then coal for steam power, combined with economic incentives to substitute technologies based on cheap coal for relatively expensive labour. We begin by examining each of these in turn.

Scientific revolution

Galileo Galilei, the Italian natural philosopher born in Pisa in 1564, has been called the 'father of the scientific method'. He practised and championed the idea that nature can come to be known through careful observation and experiments to test hypotheses. In his later life, he was placed under house arrest by Pope Urban VIII for, amongst other things, promoting the Copernican theory that the Earth revolved around the Sun. This theory was supported by Galileo's observations of moons orbiting the planet Jupiter. He also made many important contributions to physics and mechanics, including observing that falling water could be used to provide mechanical power.

Building on the work of Galileo and philosophers of the scientific method, such as Francis Bacon, the first scientific society to be set up was the Royal Society in England in 1660. This championed the use of experimental methods to improve natural knowledge. Isaac Newton, who later became president of the Royal Society, formulated his three universal laws of motion in his 1687 book on Mathematical Principles of Natural Philosophy.

These intellectual developments were not purely theoretical and there was a close interplay between scientific and practical knowledge. Later learned societies, such as the Lunar Society of Birmingham set up in the 1760s, brought together scientists, including Benjamin Franklin, Erasmus Darwin (grandfather of Charles) and Joseph Priestley, with industrialists including James Watt and Matthew Boulton, who developed and marketed the first steam engine with an external condenser. These type of interactions contributed to a setting in which new technological advances were likely to arise. However, it was the development of other political and institutional changes that created further incentives for these advances to be applied economically.

Economic developments in Britain

The so-called Glorious Revolution of 1688, in which Parliament invited William of Orange and his wife Mary to become joint monarchs of England to replace the absolutist King James II, has been identified as a 'critical juncture' by economists Acemoglu and Robinson.[1] They argue that this created a more inclusive political institutional structure, in which Parliament asserted its authority over the king, and ordinary people were given the right to petition Parliament to address grievances and perceived cases of unfairness. This, in turn, led to more inclusive economic institutions, which created incentives for investment and trade. The enforcement of property rights, including granting patents to inventors of new technologies, further stimulated innovation. This knowledge that economic gains would not be arbitrarily appropriated by the state or seized by other powerful interests thus supported the economic application of new ideas.

The importance of trade and competitive markets thus grew in Britain (following the Union of England and Scotland in 1707). Britain's naval strengths had enabled it to establish colonies or client states across the world, usually to the extreme detriment of the inhabitants of those regions. Trade in textiles from India was particularly important, creating early luxury consumption, and the production of garments in Britain from imported fabrics. It was in this context that Adam Smith wrote his classical book on the origin of the Wealth of Nations, which emphasised the economic benefits of division of labour and trading within and between nations.

These developments helped to create a business class in Britain, whose wealth depended on trade and manufacturing, rather than inherited land. This also meant that wages for workers were relatively high compared to those in other European countries.

Role of natural resources

The third key factor influencing Britain's ability to economically apply new technologies was its abundant natural resource base. In the first industrial surge from the 1770s to the 1790s, water power from fast-flowing rivers and wind power were important. In the second industrial surge, from 1829 to 1850, the use of coal for steam power was increasingly important. Abundant and easily accessible reserves of coal, particularly in the north of England, had been known for centuries, but the main practical use was the burning of coal to provide heating in homes.

Crucially for the industrial revolution, the relatively high wage levels and relative low cost of coal provided an economic incentive to invent and apply technologies that substituted energy from coal for labour.[2]

The first surge – cotton production

The cotton industry in Britain was the first to move from small-scale to industrial-scale production. Whereas the production of woollen garments relied on home-grown wool from local sheep, the production of cotton garments relied on imported fabrics, particularly from India. So, the woollen industry developed particularly in Yorkshire, whereas the cotton industry developed in Lancashire in north-west England, thanks to the proximity to the port of Liverpool for the import of cotton.

The cotton then had to be spun to produce yarn, which was then woven into garments. In the 1600s and early and mid-1700s, this was largely done in small-scale (literally) cottage industries. The invention of the 'flying shuttle' in 1733 by John Kay from Bury in Lancashire was an early mechanical device that significantly improved the productivity of weaving. In turn, this created incentives for improving the efficiency of spinning, which was then done by traditional hand-powered spinning wheels. This led to the invention of the 'spinning jenny' around 1765 by James Hargreaves, which enabled one person to operate up to 120 spinning wheels at one time.[3] The most significant breakthrough, though, was the invention of the 'spinning frame', patented in 1769 by Richard Arkwright. This was powered from a central driveshaft via a wheel and belt, and so was amenable to be powered in a large factory by water power from a flowing river. Arkwright set up his first five-storey spinning mill in Cromford in Derbyshire in 1771. This started a virtuous cycle of efficiency improvements in the technology and organisation of this mechanical production, leading to reductions in the cost of finished cotton garments. This created increasing demand for these garments, initially in Britain and then for export, and so stimulated further efficiency improvements.

The first surge – iron production

Though cotton was an important industry in its own right, the 'core input' to the first industrial surge was the production of iron. Iron had wide availability and could be used in multiple applications. Technological improvements and increasing demand led to a fall in the price of iron in the late 1700s. The first significant invention was the use of coke, derived from coal, to produce cast iron in a blast furnace by Abraham Darby in 1709 in Coalbrookdale (later renamed Ironbridge) on the River Severn in Shropshire. This enabled replacement of the use of charcoal, derived from wood, in the production of iron. As has often historically been the case, military use was a significant driver of demand. Cannon made from cast iron were needed during the Seven Years War in 1756–1763, in which Britain successfully defeated France to take possession of a number of colonial territories in North America, Africa and India. Cast iron was also used for many civilian purposes, including the development of cast iron water wheels for mills by John Smeaton, which reduced the cost per unit of available energy for water power by 70–80 per cent from 1750 to 1780.[4] Further developments were made by Henry Cort in 1783 and 1784 of rolling and 'puddling' processes for producing wrought iron, which had an even wider range of uses. Again, military uses of cast and wrought iron were important in the later Napoleonic Wars from 1803 to 1815, when the Duke of Wellington, nicknamed the 'Iron Duke', and his Prussian allies defeated Napoleon at the Battle of Waterloo.

The first surge – canal mania

The third key element of the installation period of the first surge from the 1770s to the 1790s was the expansion of transport links for moving the new textile and iron goods from production sites to market. This included both new smoother turnpike roads and the building of many new canals. Canals were built to link many rivers and population centres, particularly in the North and Midlands of England, including the Bridgewater Canal from Worsley to Manchester. Despite the expense and heavy labour needed for their building, canals came to be seen as a profitable investment, due to the massive increase in demand for cheap transport. In a pattern that was to be repeated in future surges, this led to a 'frenzy phase' of 'canal mania' in the 1790s, in which investors were persuaded in invest in increasingly unworkable new canal projects. After this bubble burst, a productive phase of investment followed.

New ways of working – rise of factory-based production

The main organisational innovation in this first surge was the rise of factory-based production in the textile sector. The technological advances by Kay, Arkwright and others enabled spinning and weaving of cotton and other textiles to take place in large water-powered factories. This facilitated the employment of large numbers of workers in these factories, with an emphasis on improving the productivity of their labour through the use of these machines and imposed work practices, such as rigorous time-keeping, that incentivised greater output. As Perez argues, the success of this new way of working in the textile sector gradually spread so that it became the 'common-sense' organising principle across other sectors.[5]

Structural crisis of adjustment

The first phase of industrialisation associated with the cotton and iron industries drove a rapid expansion of the population of cities, such as Manchester, which served as the trading hub for the textile production in the surrounding towns and villages of Lancashire. Manchester's population doubled from 70,000 in 1800 to 140,000 in 1830, indicating high rates of economic growth in these leading regions. Nevertheless, by the 1830s and early 1840s, the benefits of this first phase were becoming exhausted, leading to high rates of unemployment of even 20–30 per cent in the main industrial districts of Lancashire and Yorkshire. Though the causes of this were complex, including the downturn in military demand following the successful conclusion of the Napoleonic Wars in 1815 and global fluctuations in trade, this seemed to confirm some of the earlier fears of Luddite protestors against industrialisation.

Though they tend to be seen by history as ignorant opponents of technological progress, the original Luddites in 1811–1816 were skilled textile workers and self-employed weavers who were protesting that the introduction of spinning frames and power looms would destroy their livelihoods. After the brutal suppression of these protests, including the hanging of 17 Luddites at York in 1813, the resistance to industrial progress gradually subsided. In its place, working people began to organise for labour and political rights. This included the rise of the Chartist movement, named after the People's Charter published in 1838, which demanded the right to vote and fair representation for working men (though this was not achieved until 1867).

To overcome this downturn, the implementation of a further set of technologies was needed that would drive a new surge of industrialisation. This surge, based on the application of steam power from coal, would in fact underlie all subsequent surges to the present day, as it marked the start of the reliance on the massive energy surplus provided by fossil fuels.

The second surge – the steam engine

The development of the steam engine has a long history and the wider economic significance of its application can be dated from the launch of the first commercial and public steam-powered railway in 1829, as we shall discuss below. However, the first practical application of a steam engine was made by English engineer Thomas Newcomen in 1712 to the problem of pumping water out of mines. The key challenge for a steam engine was to create a cycle of heating and cooling of water that could drive a piston that would perform the mechanical work of pumping. Water in a cylinder could be heated by burning coal to form steam, which expands thus driving the piston up. Newcomen realised that rapidly cooling the steam in the cylinder would force it to condense back to water, creating a vacuum which would pull the piston down. Technically, the work on the down-stroke is done by the pressure of the atmosphere pushing down with the vacuum providing no counter pressure, so Newcomen's engine is sometimes called an atmospheric engine. The key point, though, is that the chemical energy of the coal is converted into mechanical energy of movement of the piston that performs useful work of pumping water. In the terminology of the time, this amounted to 'raising water by fire'.

Though this first engine had a very low efficiency (less than 1 per cent of the chemical energy of the coal was converted to perform work, with the rest lost as waste heat in the heating and cooling cycle), it met a real need. The deep mining of coal was plagued by the problem of water in the mines. By using a relatively small proportion of the coal to power his engine, Newcomen could enable mine owners to extract much more coal from their mines. The Newcomen engine thus came to be widely used for pumping water from coal mines throughout England and Scotland, as well as for pumping water from tin and copper mines.

The next decisive step in the development of the steam engine was taken by Scottish engineer James Watt. In 1765, Watt realised that the efficiency of the steam engine could be significantly increased by introducing a separate cold condenser, so that the steam in the main cylinder would remain hot throughout the cycle. A portion of this steam could be extracted and cooled in the cold condenser, which would create the vacuum needed to drive the piston down. Thanks to Watt's engineering genius, iron-maker John Wilkinson's skill in forging the cylinders and, crucially, the business acumen of Matthew Boulton, Watt's business partner, he was able to turn this idea into a machine that could be used to power a range of manufacturing processes. Boulton and Watt were able to successfully lobby Parliament for a patent for his steam engine, which was extended from 1775 to 1800. The first commercial Watt steam engine was installed in 1776 for pumping water from Bloomfield Colliery in Staffordshire. Thanks to its higher efficiency and so need for lower amounts of coal fuel, it rapidly replaced the Newcomen engine for pumping water from mines and was also used for powering blast furnaces for producing iron. Thanks to further developments by Watt, including a system for turning linear motion of the piston into rotational motion that could power machinery, the steam engine spread to other uses, including the first commercial application to a cotton mill in Manchester in 1789. This type of rotary steam power could then be applied to any manufacturing process, including iron, pottery, glass and metal production.

The second surge – the steam-powered railway

At around the same time, other engineers in competition with Watt began to experiment with the use of steam power for propulsion, either for road or rail vehicles. This would require the use of high pressure steam, in order to generate the higher rotation speeds of a crankshaft necessary to drive the wheels around to create traction in order to move the vehicle forward. For a moving vehicle, the excess steam could be expelled to the air at each stroke, though the vehicle had to carry the amount of coal fuel and water for the steam needed for the vehicle to travel a useful distance. The first high pressure steam-powered vehicle was the *Puffing Devil*, a road vehicle designed by Cornish engineer Richard Trevithick in 1801. Trevithick patented his high pressure steam engine in 1802 and it was subsequently applied to a rail locomotive, initially mainly for hauling coal on iron tracks at collieries.

The potential of a steam-powered railway to transport goods more quickly and efficiently than by canal was soon realised. This led to a series of trials in 1829 for a locomotive for the planned 35-mile railway to transport goods from the port of Liverpool to the industrial city of Manchester in north-west England. The winner of the trials was *The Rocket*, designed by father-and-son team George and Robert Stephenson, which was able to pull 13 tons at an average speed of 12 miles per hour and a top speed of 30 miles per hour. The Liverpool–Manchester railway was opened in 1830, and quickly became successful (despite the accidental death of local politician William Huskisson on the first journey). One feature that was surprising to some of the engineering pioneers was that passenger journeys by rail, particularly for leisure travel, were equally as popular as the transport of goods.

The building of railways rapidly expanded, with over 10,000 miles of track built in Britain by 1860, which would carry around 300 million passengers by 1870. The transport of freight goods was still the most economically significant use. As the scale of the railway network increased and the efficiency of steam engines improved, the price of transporting goods fell, and so demand increased to over

100 million train miles of freight, including 6.2 million tons of coal, carried by 1880.[6] The potential profits to be made by these new railway lines resulted in waves of investment in the 1830s and 1840s. Almost inevitably, this led to a frenzy of speculative investment in uneconomic as well as economic construction projects, with outbursts of 'railway mania' in 1834–1837 and 1844–1847. After the bursting of these bubbles, a phase of more productive investment in the expansion of the railways occurred in the 1850s and 1860s.

The second surge – positive feedbacks and complementarities

The rapid economic expansion that is typically associated with the industrial revolution thus occurred in the 1850s and 1860s, This was thanks to synergies between the use of stationary steam power in factories for cotton, iron, pottery and other goods, the use of motive steam power for railways to transport goods and people across the country, and the development of machine tool production to make engines, tracks, looms and other tools. This also enabled the diffusion of complementary innovations, such as the electric telegraph. Following the first commercial electric telegraph, designed by William Cooke and Charles Wheatstone in 1837, this was rapidly installed alongside new railway tracks in the 1840s, creating the world's first long-distance rapid communications system.

A further application of steam power to transport was the design and building of iron-hulled steam ships by Brunel and other engineers. However, it was not until the 1880s that the tonnage carried by steam ships registered in Britain exceeded that carried by sailing ships. The study of the transition from sailing ships to steam ships highlighted the key niches in which the new technology could flourish, such as reliable transatlantic crossing, and the influence of wider social and cultural 'landscape' changes that helped to stimulate this transition.[7]

The core inputs for this economic expansion were coal and iron ore, which could both be mined relatively cheaply in Britain. As the demand for coal and iron soared and the railways enabled their cheap transportation, their prices dropped. For example, the price of coal in Manchester dropped from 16 shillings per ton in 1800 to 6 shillings per ton in 1850.[8] In turn, the availability of cheap coal and iron stimulated their widespread use across other industries, and the innovation of complementary technologies to take advantage of this.

The role of railroads in the United States

In the United States of America at this time, the Northern and Western states had already begun along the road of industrialisation, whilst the economies of the Southern States were dominated by agricultural, particularly cotton, production, enabled by large numbers of slave labour. Following the victory of the Northern Union over the Southern Confederacy in the American Civil War in 1865, slavery was abolished throughout the United States. This enabled the spread of the railways and capital-intensive mechanisation of industry, based on the complementary growth of the US coal and iron industries. By 1869, the first transcontinental railway was opened, and over 5,000 miles of new track was added each year in the early 1870s. This facilitated the expansion of the population and the creation of a huge national market for goods and services, which enabled the US to overtake Britain in terms of productivity levels, to become the largest economy in the world.

Structural crisis of adjustment

In Britain, though, by the late 1870s and 1880s, the rapid gains in productivity from the new industries based on steam power and the spread of the railways had begun to decline. The maturity of these industries led to declining profitability, intensified competition and diminishing opportunities for new profitable investment, resulting in high levels of unemployment in the 1880s in Britain. This would then set the stage for a third industrial surge, which would come to fruition in the 1890s and early 1900s, but this time led by the United States of America.

New ways of working and a new energy basis

This second surge of industrialisation, particularly from the 1850s to the early 1870s, first in Britain and then in continental Europe and the US, was the first sustained period of modern economic growth. This was enabled by the synergies and positive feedbacks between the application of steam power in factories and the spread of railways providing cheap transport of goods and people. As we saw at the beginning of this chapter, this was facilitated by complementary developments in scientific, economic and political spheres, together with the availability of natural resources. This led to a process of coevolution between technologies, institutions and business strategies, and together with natural ecosystems.

Though the development of steam power largely progressed through the trial-and-error experimentation of engineers such as Newcomen, Watt and Trevithick, they were operating in a scientific and economic context that supported the dissemination of knowledge and its economic application. The development of political and economic institutions, such as stable property rights, patents for inventions, limited liability companies and banks and other forms of financial support, were important to this application of knowledge.

In energy terms, the harnessing of chemical energy in coal to provide mechanical energy for powering machines and kinetic energy of motion marked a turning point in the creation of wealth in human history (literally, as shown in Figure 5.1). Up to this point, human appropriation of energy was limited by the flows of energy from the Sun, captured by plants. The main energy sources for heating and for mechanical work were thus wood from forests and draft animals, such as oxen, fed from pastures. These were limited by land area available and the high cost of transporting wood over long distances. The use of coal as a primary energy source, and subsequently oil and gas, enabled these limitations to be overcome. The

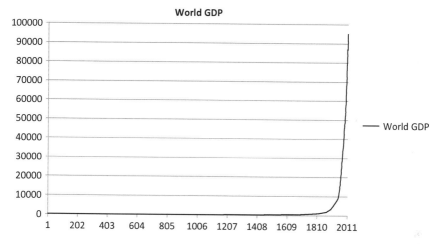

FIGURE 5.1 World GDP, Year 1–2011 (GDP in 2011 International $).
Source: Data from Maddison (2010).

chemical energy in coal comes from solar energy buried and concentrated by pressure over millions of years to form a 'subterranean forest'.[9] The ability to release this energy to provide mechanical work and motion thus liberated humanity from land-constrained energy sources, and enabled an exponential expansion in the quantity and quality of goods and services that could be supplied to meet people's demands for nutrition, mobility, communication, clothing and other services.

Costs of energy expansion

However, this expansion is not without costs, as we shall discuss in more detail in later chapters. The first concern is that these concentrated energy sources, or at least the most easily available of these sources, will peak and start to decline. The second concern is the ability of the Earth's atmosphere and biosphere to absorb the waste products of these energy sources. The burning of coal and other fossil fuels produces carbon dioxide and other greenhouse gases that are now accumulating in the atmosphere and in the ocean, as the rate of release is higher than their natural rate of breakdown by the biosphere. Of course, the latter concern was not known to the early industrial pioneers.

Interesting, though, the question of the continued availability of coal was already being asked by English economist William Stanley Jevons in his 1865 book *The Coal Question*. Jevons noted that the development and application of the steam engine by Watt and others had greatly increased the efficiency of producing useful work from the chemical energy in coal. However, this increase in efficiency had not led to a decrease in the amount of coal being used. This was because the increase in efficiency, and associated decrease in cost, of the use of coal stimulated many more and different uses of steam power, as we have seen. This then increased the demand for coal. As Jevons put it, 'It is wholly a confusion of ideas to suppose that the economical use of fuel is equivalent to a diminished consumption. The very contrary is the truth.'[10] Jevons was concerned that this ever-increasing demand for coal, coupled with limited coal reserves in Britain, would eventually lead to shortages and higher extraction costs, and the loss of Britain's competitive advantage in manufacturing and transport.

This highlights two key points that environmental and ecological economists have argued over ever since. First, the extent to which reductions in primary energy demand due to improvements in the efficiency of energy conversion are taken back due to increases in the demand for energy services, as efficiency improvements lead to a reduction in the cost of those services. This is sometimes referred to as the 'Jevons paradox', though it is a question of the scale of this take-back or 'rebound' resulting from different types of energy efficiency improvement. Second, the question of the energy that needs to be invested in order to give a useful energy return. As Jevons realised, easily accessible sources of fossil energy will be extracted first, which give a high energy surplus. As these reserves are used, more and more inaccessible sources will need to be extracted, which will have higher costs in monetary terms and in the amount of energy that needs to be invested to extract them, thus reducing the energy surplus produced.

We shall return to these questions in later chapters. In the next chapter, however, we continue our historical story with the next industrial surge, based on steel, heavy engineering and electrification, which was led by the US.

Notes

1 Acemoglu and Robinson (2012).
2 Allen (2009). We discuss these economic incentives for the industrial revolution in more detail in Chapter 9.
3 Osborne (2013).
4 Tylecote (1992).
5 Perez (2002).
6 Freeman and Louçã (2001), p. 191.
7 Geels (2002).
8 Von Tunzelmann (1978).
9 Sieferle (1982/2001).
10 Jevons (1865), VII.3.

6

ELECTRIFICATION AND THE RISE OF OIL

Introduction

This chapter looks at the installation and deployment of the third surge from the 1870s to the 1900s associated with the initial implementation of electrification and the rise of the steel industry, both of which contributed to increasing use of coal. It goes on to examine the installation of the fourth surge from 1908, relating to the rise of mass production and the start of the automobile industry, which was ended by the Wall Street Crash in 1929 and the Great Depression in the 1930s.

A new global surge

By the 1880s, the surge of new industrial activity associated with the application of steam engines, powered by coal, to railways and iron production in the UK and then other Western countries was reaching diminishing returns. However, the installation period of the next surge had already begun. This would be based particularly on the development of the steel industry, heavy engineering and widespread electrification to provide power to homes and businesses. The production of steel was still based on steam power from coal, but the development of the new Bessemer process significantly increased the efficiency of its production. As the cost of steel reduced, its desirable qualities of strength combined with workability meant that steel was able to replace iron for many uses. The spread of electrification meant the rise of a new efficient and flexible energy carrier, which enabled a wide range of end uses. Again, these developments required interacting changes in institutional rules and also stimulated organisational change through the rise of new large, hierarchical companies in these sectors. The golden age of these developments would be led by the US and Germany in the 1890s and early 1900s.

As this surge reached maturity by the time of the First World War, 1914–1918, the installation phase of the following surge had begun, again led by the US. This would be based on new mass production processes, the automobile as a key enabling technology and a new primary energy source, oil. However, the golden age of these developments, linked to social changes including suburbanisation and the rise of mass consumption, would not occur until the 1950s and 1960s.

The third surge – electrification

Like other developments that we have examined, the spread of electricity from the 1890s rests on a long history of earlier scientific and technological advances. Like the development of steam power, the commercial application of electricity relied on a combination of scientific knowledge, economic incentives embedded in markets and institutions, and dependence on the use of natural resources. The natural phenomena of electricity, in the form of lightning and static electricity from amber, and magnetism, that certain ores of iron would exhibit a force of attracting and repelling similar ores, were known back in antiquity. However, the understanding and practical application of these phenomena dates largely from the 1700s onwards. The exception was the use of a magnetic compass, influenced by the Earth's natural magnetic field, as a directional aid by the Song Dynasty Chinese in the eleventh century and in Western Europe and Persia by the thirteenth century.

Early experimenters included American natural philosopher (and later political theorist and statesman) Benjamin Franklin, who showed in 1752 the equivalence of electricity produced in lightning with the static electricity that other experimenters had produced by the frictional rubbing of certain substances, such as amber. The first source of a continuous electric current was the electric battery, consisting of alternating zinc and copper discs separated by cardboard or cloth soaked in saltwater, invented by Italian Alessandro Volta in 1800. The link between electricity and magnetism was first understood in a series of experiments by Hans Christian Oersted in Denmark, André-Marie Ampère in France and Michael Faraday, working at the Royal Institution in London. In 1832, Faraday produced the first electric generator, by showing that a moving magnet would produce an electric current in an external circuit. This enables mechanical energy of motion to be converted into electrical energy. As already noted, the first commercial application of electro-magnetism was the electric telegraph for long-distance communication, invented in 1837 and frequently installed alongside new railway lines. Building on the work of other inventors, the first practical electric motors were invented in America by Frank Julian Sprague in 1886 for direct current (DC) and by Serbian émigré Nikola Tesla in 1887 for alternating current (AC). These performed the reverse operation of Faraday's electric generator by converting electrical energy into mechanical energy. This enables a complete electrical system to be created in which distantly generated electricity could be transmitted to an end user where electric motors could be used to power devices or machinery.

As with the role that Matthew Boulton played in the application of the steam engine, the commercialisation of electricity required business as well as technical skills, exemplified by two American pioneers of electrification, Thomas Edison and George Westinghouse. Though Edison is best remembered for his important inventive contributions, including the phonograph (for reproducing sound), the motion picture camera and the light bulb, he also pioneered new ways of working and new business strategies. In 1876, Edison set up one of the world's first industrial research labs in Menlo Park in New Jersey. Over the next ten years, he employed hundreds of engineers and technicians, including Sprague and Tesla, to innovate and demonstrate many new technologies. After his invention of the phonograph helped to make his name, one of his early financial backers was the financier J.P. Morgan.

Edison was the first to have a vision of a complete electrical system, which he set up in New York in 1882, powered by the Pearl Street coal-fired power station in lower Manhattan. This supplied power to 59 houses, which were then lit up using his newly invented light bulb, and so competing against the then-dominant gas lighting. The system soon included an electric meter, so that homes could be charged based on how many kilowatt-hours of electricity they had consumed. He pushed the benefits of a centralised 'utility' company that end users could connect to, rather than a decentralised system where users would generate their own electricity. This new system required a new institutional framework to regulate it and Edison's lawyer, Samuel Insull, helped to push through regulatory and trade standards decisions in support of his system.[1] Edison didn't have things all his own way, though, and what has come to be known as 'the battle of the currents' ensued.[2]

Edison's system was based on direct current (DC). He argued that this was safer, as it ran at a lower voltage (110 volts), which was suitable for his high resistance incandescent lamps. However, direct current suffers significant losses if it is transported over long distances. His competitor, inventor and entrepreneur George Westinghouse, pioneered a system based on alternating current (AC). This has the advantage that the voltage can easily be transformed up to higher voltages for transmission with fewer losses, before being transformed back down to lower voltages for end use. In 1888 and 1889, the battle of the currents between DC and AC took place in newspapers and public debates, as well as in the pages of technical journals. Most bizarrely, this included a campaign by engineer Harold Brown, partly funded by Edison's company, to associate a proposed new form of execution, the electric chair, with Westinghouse's AC system. In the end, though, the AC system won out, thanks to the ability to transport electricity over large distances and Tesla's efficient electric motor for AC. In 1896, a long-distance transmission system was set up to transmit electricity from a hydro-electric power station at Niagara Falls to supply the city of Buffalo, 20 miles away. This was soon extended to New York City, becoming the start of continental electricity grids.

Though all electricity systems around the world now run on AC, in recent years advocates of DC are making a comeback. Precision electric devices such as computers are designed to run on DC (and hence require the heavy black box on the power cable to convert from AC to DC). Small-scale generation devices, such as solar PV cells, produce DC, and so there are potential synergies there. High voltage DC power cables have also been proposed for long-distance transmission of solar energy, such as from North Africa to Europe.

Although the potential benefits of the new electricity systems were clear, their economic impacts took longer to realise. Again, this reflects the time taken for technological and institutional systems designed to fit the previous techno-economic paradigm to adapt to realise the benefits of the new system. In particular, factories designed to use steam power operated with a central rotating shaft, from which all machines drew their power. At first, large electric motors simply replaced steam engines in this centralised factory design. However, factory managers began to realise that the flexibility of electrical supply meant this could be replaced by wires and smaller electric motors designed to meet the power requirements of different machines. This meant that space could be utilised more effectively and portable power tools could be used. This improved both productivity and product quality in industries such as textiles and printing. However, economic historian Paul David has argued that it took at least 20 years, until the 1920s, for the widespread diffusion of these new electrified factory systems and so for the benefits to show up in productivity statistics.[3] David was noting the parallel with the arguments then current in 1990 that the economic benefits of the spread of computing power were not being realised. He quoted a remark made by economist Robert Solow that 'we see the computer age everywhere but in the productivity statistics'.[4] In a similar way, the productivity benefits of computers were not widely seen until systems, based on the previous mass production paradigm and hierarchical organisational structures, were redesigned to take advantage of the flexibility and new applications that could be harnessed by the use of computers. This argument supports Freeman and Perez's view of the emergence of a new techno-economic paradigm as a long-drawn-out and contingent process.

The third surge – steel industry

The production of cast or wrought iron was already important in the second surge for use in railways and other engineering uses. However, the usefulness of iron was particularly sensitive to the level of carbon introduced into the final product – too much carbon made it brittle, too little carbon meant it lacked strength. It was recognised that steel, which is iron with around 1.5 to 2 per cent carbon, is the variant with the right combination of strength and suppleness for many applications, but this was difficult to produce in large quantities. In 1856, Englishman Henry Bessemer invented a process for the mass production of high quality steel. This process involved taking high carbon pig iron from a coal-fired blast furnace and blowing air through it to reduce the carbon content (releasing carbon dioxide) and remove other impurities. Bessemer patented the process and set up his own steel works in Sheffield to mass produce steel. In 1872, Andrew Carnegie, who was to become one of the main American steel magnates, visited the Bessemer plant and was so impressed that he secured finance for the building of the new Edgar Thomson Steel Works in Pittsburgh to apply the process under licence. This Steel Works, opening in 1875, initially supplied rails for the Pennsylvania Railroad, and has been in continuous operation ever since to the present. The subsequent expansion of steel production in the US dramatically reduced the cost of steel rails from $107 per ton in 1870 to $18 per ton in 1898. This cost reduction, together with its higher performance compared to iron, enabled steel to be used not only for railways, but also for shipbuilding, machine tools, construction and other engineering uses. Carnegie was one of the leaders of this expansion, combining technological innovation with organisational innovation in his management processes to improve efficiency and productivity of operation (whilst suppressing workers' rights).[5] This enabled him to expand and diversify his enterprises, making him one of the richest men in the world by 1900.

Expansion of coal use

Though the third economic surge was not associated with a new source of energy, it was enabled by a rapid expansion in use of the dominant energy source of coal, which, partly thanks to earlier technological innovations, was relatively cheap to mine. The burning of coal quickly became the main source of energy for the generation of electricity in large centralised power plants. This is essentially based on a scale-up of Faraday's electric generator. In a coal-fired power plant, coal is burned in air at high temperature to turn pure water into steam, releasing large amounts of CO_2. This superheated steam then turns a turbine, which is connected to a dynamo consisting of a moving magnet that generates an electrical current in coils of wire. The magnet moves at 50 revolutions per second to create alternating current at 50 Hertz. To complete the cycle, the steam has to be condensed back into pure water. This is usually done with a secondary cooling water system, including the familiar cooling towers releasing water vapour into the atmosphere. In this process, only about 40 per cent of the primary energy (exergy) content of the coal is converted into electricity with the rest being lost as waste heat in the water vapour. Coal was a more flexible energy source than hydro power, as it could be shipped or transported by rail to where it was needed. However, economies of scale associated with large power plants and the relative cheapness of transmitting electricity by wire (despite additional transmission losses) compared to transporting coal led to the familiar electricity network consisting of large centralised generation and high voltage AC transmission networks. The economic value of electricity as a high quality and flexible energy carrier meant that the business of generation was economic despite technical energy efficiencies of 40 per cent or less.[6] Of course, the side effect of the release of large amounts of CO_2 and its contribution to global climate change was not known at this time.[7]

Coal was also the main source of energy for steel production. Coal in the form of coke (which had previously had its water content burned off) was used as the fuel in blast furnaces for smelting iron from iron ore and for blowing the high pressure air through the pig iron to produce steel. This also led to release of CO_2.

International trade and leadership

Britain and other European countries, particularly Germany, benefited from the expansion of free trade and economically liberal laissez-faire policies of governments that encouraged this in the deployment phase of the second surge from 1848 to the 1870s. However, as diminishing returns began to set in to the expansion of industries based on steam power and iron, an economic depression occurred in the 1870s. This helped to stimulate the rise of both socialist and nationalist ideologies and a shift away from economically liberal policies.[8] Global economic leadership thus passed to the US, which was experiencing high rates of economic growth associated with investment in electrification, steel production, railroads and related industries.

In Britain, financial institutions at this time began to increasingly focus on foreign investment in Britain's overseas colonies, both in plantation and mining activities and in investment in infrastructure and services. This again led to financial bubbles, the most famous of which almost led to the bankruptcy of Barings Bank in London, after it had over-invested in developing infrastructure for meat and wheat exports from Argentina.[9] This type of foreign investment reduced the level of domestic investment in the electrification, steel and engineering technologies of the third surge, despite Britain's technological leadership in some of these areas. Unlike the US and Germany, Britain also failed to invest in the development of new institutions of education and training relevant to these new technologies.

This process of coevolution of technological, institutional and business organisation innovations created positive feedbacks of economic growth and development in the late installation and deployment phases of the third surge up to 1918. This resulted in Britain being surpassed in terms of manufacturing output by the US by 1900 and by Germany by 1913. The resulting rapid expansion in per capita incomes between rich and poor countries opened up the division between rich industrialised countries and poor countries with agriculture and commodity-based economies that is still with us today.

War and revolution

This economic expansion in Europe helped to exacerbate political tensions that had previously largely been contained through a complex system of international alliances. Following the unification of the German state in 1871 after victory in the Franco-Prussian War, Germany devoted a significant amount of its new economic resources to the expansion of its military, particularly its Navy. This led to an arms race with Britain and its dominant Royal Navy, with other European nations following suit by expanding their military spending. After alliances had been triggered by the assassination of Archduke Franz Ferdinand of Austria by a Serbian nationalist, the 'Great War', first major war of the mechanised era, broke out in 1914. Known as the First World War (1914–1918) after also bringing in the United States and Japan, this led to the death of over 9 million military personnel and over 7 million civilians, as new military technologies of artillery, machine guns, tanks and mustard gas led to prolonged trench warfare.

The war also precipitated the Russian Revolution of 1917, which enabled the Bolshevik socialists led by Vladimir Lenin to seize power and implement a form of communism inspired by the ideas of Karl Marx and Friedrich Engels.

The heavy 'reparations' payments imposed on Germany by the victorious Allies helped to create further economic instability in the 1920s. As British economist John Maynard Keynes had predicted in his 1920 book on *The Economic Consequences of the Peace*, Germany was unable to keep up these payments, except by printing money which led to hyper-inflation in 1923.[10] This eventually resulted in the collapse of the German economy in 1931, precipitating the rise to power of the Fascist Nazi Party under Adolf Hitler. This led to the Second World War (1939–1945), in which over 50 million military personnel and civilians died, enabled by technological advances and political ideologies.

The fourth surge – automobile industry and mass production

As discussed in Chapter 5, the invention and development of the steam engine enabled the conversion of heat energy into mechanical energy, which could be used to power machinery or for propulsion. Though the steam engine was applied to individual road vehicles, the solid fuel (coal) was heavy and the vehicle required regular refilling with water. The first internal combustion engine which burned gas or liquid fuel was invented by Belgian Jean Étienne Lenoir in 1860. By 1863, he had fitted this to the first automobile, the Hippomobile, fuelled by hydrogen gas. Building on this, the four-stroke compression ignition engine, fuelled by petroleum gas, was developed by German Nikolaus Otto in 1876. The more efficient diesel engine, in which liquid gasoline fuel is injected into the engine and ignited by high temperature of compressed air, was developed by German Rudolf Diesel in 1892. By the turn of the twentieth century, gasoline-powered vehicles were competing with steam-powered and electric-powered vehicles, both in Europe and in the US.

A significant breakthrough occurred with the first production of the Ford Model-T motor car in 1908. This was the first product to be made on a mass production assembly line, with workers specialising in particular tasks and fitting interchangeable parts, implementing Adam Smith's division of labour and leading to big improvements in productivity. Further productivity improvements followed from Ford's introduction of the moving assembly line in 1913. This enabled a rapid reduction in the cost of production, with the price of the car to consumers falling by over 50 per cent to $360 in 1916. The low cost, robustness and longer operating range of the Model-T proved attractive to consumers, and 15 million of these cars were sold by 1927. A further innovation made by Ford was to increase the wages of his workers to $5 per day in 1914, though this was partly to offset turn-over of labour due to the unpleasant nature of production line working. This meant that the cost of owning a car came within the scope of a manual worker for the first time, setting off the positive feedback between production and consumption which categorised this surge of economic growth.

Following the success of this production process, other manufacturers quickly followed suit and similarly scaled up production, introducing incremental process improvements, but no radical change in production or design. This resulted in a small number of firms, in this case Ford, GM and Chrysler, out-competing other firms and leading the industry for the next 50 years, based on the shared dominant design of the gasoline-powered motor car.[11] The infrastructure of refuelling stations, repair garages, second-hand car dealers, together with the regulatory and institutional frameworks of highway rules, safety and fuel economy standards, speed limits, etc., built up to support the vehicles helped to reinforce this 'lock-in' of the entire gasoline-based personal transport system. User practices of driving to work and for leisure also evolved together with the technological and institutional system. The American dominance of car production was first threatened in the 1980s, as Toyota and other Japanese manufacturers introduced a 'lean production' system, which was based on just-in-time production, continuous process improvement and elimination of waste. The US car industry recovered somewhat in the 1990s, partly thanks to the popularity of larger SUVs (sports or suburban utility vehicles) which were exempt from the Corporate Average Fuel Economy (CAFE) standards. However, following the 2008 financial crisis, both GM and Chrysler had to be bailed out by the US government, under the Troubled Asset Relief Program (TARP), to avoid bankruptcy.

Cheap oil in the fourth surge

Though drilling for oil and its transport by pipeline began in Imperial Russia in the 1840s, it was the first deep reservoir oil drilling at the Drake Well in Pennsylvania, US, in 1859 that marked the first modern oil well. However, in order to be useful for conversion to motive power in engines, crude oil needed to be refined into its constituent parts, through a process known as cracking which applies high tempera-tures and pressures to the oil. The first successful commercial cracking process was developed by industrial chemist William Burton in Standard Oil's research laboratory in Indiana and came into operation in 1913. Thus, by happy coincidence, the pro-cess for commercial production of gasoline, which quickly became the dominant fuel for the motor car, was inaugurated in the same year as Ford's moving assembly line for the car's production. The reduction in cost and improvement in performance of vehicles thanks to these two technological and organisational advances, together with a latent demand for personal mobility, helped to expand car ownership to nearly 50 per cent of American households by 1930.

Despite the Great Depression in the 1930s, major collaborative R&D pro-grammes between oil companies and other process companies resulted in further technological improvements leading to the fluid catalytic cracking process. Thus, a combination of economies of scale and technological innovation significantly reduced the cost of refined oil and the expansion of supply of gasoline for cars, diesel for trucks and rail, and kerosene for aviation fuel.

Haber–Bosch process

The other key technological development at this time in terms of energy use and impact on human wellbeing and the environment was the Haber–Bosch process for synthesising nitrogen-based fertilisers, such as ammonia. This has been called 'the most important invention of the twentieth century'.[12] Nitrogen is a key nutrient required for healthy soils and plant growth, but apart from legumes such as peas, very few plants help to return nitrogen to the soil. Natural sources of nitrogen were saltpetre and guano (bird droppings) which required expensive transportation. Though nitrogen makes up 80 per cent of air, it is very stable and so difficult to combine into useful compounds. In 1909, German industrial chemists Fritz Haber and Carl Bosch invented and developed a process for producing ammonia at industrial scales by combining nitrogen and hydrogen using a chemical catalyst at high temperatures and pressures. Though this process is very energy intensive, the value resulting from soil fertilised with ammonia offset these costs by significantly raising crop yields. In turn, this improved diets and health in industrialised countries where the process was applied, and helped to enable the large increases in population in the twentieth century. It is estimated that the Haber–Bosch process still uses about 1–2 per cent of the world's annual energy supply.

Great Depression

After the trauma of the First World War (1914–1918), the 1920s was generally a period of optimism, known in the US as the 'Roaring Twenties'. This reflected the prosperity created by the deployment of the third surge technologies of elec-trification, steel and heavy engineering, and the rise of the financial sector in the US, with unemployment rates reducing to around 4 per cent by 1926. This com-bined with the growing but uneven installation of automobile manufacturing and the oil industry. With the economy growing and confidence spreading, Americans began to take on more private debt, partly to fuel a housing and construction boom. The value of traded companies soared on US stock exchanges, creating a financial bubble. This all came crashing down in 1929, starting with the Wall Street Crash on the New York stock exchange on 29 October 1929, as investors rushed to sell overvalued shares. This is generally seen as precipitating the Great Depression of the 1930s, though the causes of the depression continue to be debated.

Following Perez, the period of structural adjustment between the third and fourth surges played a key role in the crisis. The demands for automobiles and housing led to booms in these industries, but the infrastructure and institutions were not yet in place for the widespread deployment of these new industries. For example, in the early 1920s, there were popular protests about the inadequacy of the roads network, which were unable to cope with the impacts of a housing boom in places such as Florida. This resulted in overcapacity in the auto production industry, with the total output of vehicles falling from a high of 4.8 million in 1929 to 2.8 million in 1930, and down to 1.1 million vehicles by 1932. As confidence in the future of the US economy collapsed following the Wall Street Crash, the value of worldwide economic output fell by around 15 per cent between 1929 and 1932, as both manufacturing and service sectors experienced severe downturns, with US unemployment levels rising to around 25 per cent by 1933. Partly thanks to a reduction in previous high levels of lending to firms in Europe and elsewhere by US banks, the depression spread globally. In Germany, this combined with domestic factors relating to the paying off of reparations from the First World War to create a severe economic depression, with unemployment levels soaring to 13 million by 1932. As we shall see, the very different responses in the two countries had huge social and economic consequences.

Economists have debated the relative roles of supply and demand in leading to the depression ever since. The mostly widely accepted explanation for the overall rapid decline in economic output is that given by British economist Keynes in his 1936 book, *The General Theory of Employment, Interest and Money*.[13] Keynes argued that it is the lack of aggregate demand in the economy that gives rise to recessions. Unlike earlier classical economists who believed that competitive markets would always lead to full employment, as labour would be willing to accept jobs at lower wages, Keynes argued that a destructive spiral can occur as a reduction in confidence leads to a decrease in spending in the economy on goods and services, further reducing private investment and output. Thus, Keynes stressed the consumption side of the economy as mainly driving the production side. His remedy for the resulting recession is action by central banks to reduce long-term interest rates and by government to stimulate the economy by public spending, to counteract the downturn in private spending.

However, it is also accepted that monetary factors played a significant role in the depression. American economist Irving Fisher argued that the US had entered a debt–deflation spiral.[14] In the run-up to the crash, households and firms had taken on higher levels of private debt, creating a financial bubble. When the crash came and investors began to sell, loans could not be repaid, leading to large-scale default on these debts. As worried depositors then sought to withdraw their deposits from banks, this led to bank runs and multiple bank failures, with losses of $7 billion by 1933. As asset prices fell, this led to generalised price decreases, known as deflation. This leads firms to reduce their investment and households to reduce consumption, in the expectation that goods will become cheaper in the future. Banks also sought to build up their capital reserves and reduce lending, intensifying deflationary pressures. The fall in prices also made debts harder to pay off. As Perez has emphasised, this type of credit crunch is clearly reminiscent of what happened to economies of industrialised countries following the financial crisis in 2008.[15]

The response to the Great Depression in the US by the administration of President Franklin D. Roosevelt and the US Congress, referred to as the New Deal, incorporated aspects of both monetary and demand-side responses. The immediate response in 1933 focused on restoring confidence in the banking sector by passing Banking Acts that created federal deposit insurance schemes. The Glass–Steagall Act of 1933 also separated commercial 'high-street' banks from investment banks. This separation was maintained until the Act was dissolved under the Clinton administration in 1999, which some have argued was a contributory factor to the 2008 financial crisis. Between 1935 and 1938, the New Deal was expanded to promote public spending on infrastructure programmes that helped to create jobs, thus following the Keynesian prescription. This led to the construction of thousands of public buildings, roads, schools and other facilities, as well as other job-creating and welfare-enhancing projects. New Deal legislation also created social security programmes, including unemployment relief, and enhanced trade union rights, ensuring the ability of workers to negotiate fairer pay deals. However, it was only with a massive increase in public spending and support for private sector employment linked with the entry of the US into the Second World War in 1941 that the Great Depression was finally ended.

The response to the Great Depression in Germany, of course, took a much darker turn. With the support of industrialists such as steel and arms manufacturer Krupp, Hitler came to power in 1933. He soon seized complete dictatorial powers, and set about a process of investing in arms production to achieve the rearmament of Germany. Together with building the Autobahn road network and car manufacturing through the Volkswagen state-owned company, and aided by industry-backed finance bonds, this ended the depression through the creation of a military–industrial state. Inevitably, this led to the Second World War, as Britain (from 1939), the US (from 1941) and their allies fought to restore freedom and democracy to Europe.

As we know, thankfully, Nazi despotism in Europe and Japanese imperialism in Asia were defeated by the Allies by 1945. A series of far-sighted decisions enabled Germany and Japan not only to return to peaceful and democratic governments after the war, but also to prosper economically by taking a full part in the deployment phase of the fourth surge. To aid economic recovery, the US gave over \$12 billion to rebuild the shattered economies of Britain, France, West Germany and other Western European countries through the Marshall Plan, beginning in 1948. This helped to pay for the import of raw materials, products and machinery from the US, as well as the provision of technical assistance to improve the productivity of European firms. It also included the creation of funds in local currency to support reindustrialisation. Crucially, this included the founding of the KfW bank in Germany (literally, Reconstruction Credit Institute), which is still providing support for German industry to this day.[16] The KfW bank managed a special fund, which provided low-cost loans, initially to housing and energy supply firms. As these loans were repaid, they were recycled to provide a revolving fund for further investment in old and new industries, which had provided low-interest loans worth up to DM 140 billion by 1995. Through this fund, together with an export financing facility that financed and profited from the development of mainly small and medium-sized manufacturing firms that successfully exported goods all over the world, the KfW bank helped to finance the reconstruction and development of the German economy to now be one of the strongest economies in the world.

Deployment of the fourth surge

As we have seen, the expansion of industrial activity associated with the New Deal and the mobilisation for the Second World War overcame the Great Depression that followed the financial crash of 1929. The installation phase of the fourth surge around automobiles, oil as a cheap energy source and organisational innovation of mass production techniques had already taken place, and dominant designs had been selected. Together with the rebuilding of European and Japanese economies associated with the Marshall Plan and other refinancing approaches, this meant that the stage was set for the deployment of these technologies and innovations to realise a new golden age of economic growth starting in the 1950s. This would be particularly associated with the rise of mass consumption in industrialised countries, as we shall discuss in the next chapter.

Notes

1 Freeman and Louçã (2001), p. 229.
2 Hughes (1983).
3 David (1990).
4 Solow (1987).
5 Chandler (1990).
6 Schurr and Netschert (1960).
7 This is known as a negative externality in mainstream economics.
8 Freeman and Louçã (2001), pp. 218–219.
9 Ironically, Barings Bank was brought down just over 100 years later in 1995 by speculative investments in financial derivatives by 'rogue trader' Nick Leeson with inadequate management supervision.
10 Keynes (1919).
11 Utterback and Abernathy (1975) introduced the concept of a 'dominant design', i.e. a standard that becomes widely adopted after a period of innovation and competition.
12 Smil (1999).
13 Keynes (1936).
14 Fisher (1933).
15 Perez (2016).
16 KfW (no date).

7

RISE OF THE CONSUMER SOCIETY AND THE ICT REVOLUTION

Introduction

So far, we have identified four constellations of new technologies, institutions, strategies and practices that gave rise to surges of economic growth, often beginning in one or two leading countries before diffusing to other industrialised countries. The first industrial revolution, associated with mechanisation of the cotton industry and water power, began in around 1770 in Britain. The second surge, associated with steam power and the rise of the railways, began in 1829 in Britain and quickly spread to continental Europe and the US. The third surge, associated with the generation of electricity, the production of steel and heavy engineering, began in 1875 in the US and Germany. The fourth surge, associated with the use of oil, automobiles and mass production techniques, began in 1908 in the US with the opening of Ford's Model-T car production plant in Detroit. As we have seen, though, in each case, the surge of economic growth only materialised 20 or 30 years later in a deployment phase that followed a long installation phase and the bursting of a speculative financial bubble.

In this chapter, we discuss the deployment of the fourth surge associated with the post-war golden age of the 1950s and 1960s, and the installation of the fifth surge, associated with new information and communications technologies (ICT), from the early 1970s, cumulating in the dot-com bubble of the early 2000s and the global financial crisis of 2008.

Though the earlier deployment phases were associated with new consumer demands, such as cotton clothing, leisure travel using railways and electric lighting, their primary economic impacts were in significantly improving the production and supply of goods. For example, railways enabled the easier and more rapid transportation of freight, and electrification enabled the use of electric motors which improved the efficiency of many production processes. The deployment of the fourth surge, in contrast, ushered in a new era of mass consumption, with new consumer practices arising to take advantage of the mass production of goods and the ease of transportation of both people and freight enabled by the use of cheap oil. This enabled a process of suburbanisation, with high levels of investment in the building of housing and roads for a burgeoning middle class to live in new suburban communities. This was supported by action by the state, including the setting up of welfare state provisions, such as health care and unemployment benefits, that smoothed the path to new employment opportunities.

The rise of mass consumption

As we saw in the last chapter, the installation phase of the mass production and automobile paradigm occurred first in the US in the 1920s. This was ended by the crash of a speculative bubble and the Great Depression in the 1930s. Though President Roosevelt's New Deal was important in stimulating the levels of investment needed to ease the depression, it was not until the investment associated with the production of armaments and oil-powered vehicles for the Second World War that the US finally escaped from the depression. Unlike Europe, which was physically and mentally exhausted by the end of war, the US had not suffered internally from significant bombing impacts and had lost a far smaller proportion of its population to death or injury. In the late 1940s and 1950s, the US was thus able to reorient its wartime production to the mass production of cars, electrical equipment and other consumer durables. The purchase of these products was facilitated by the rise of consumer credit innovations, such as hire purchase or instalment plans, which spread payments over a longer period.

This created a positive feedback loop between the expansion of demand for goods and services and the production and supply of those goods and services in the US, leading to a period of sustained high economic growth. This also stimulated high levels of energy use, both in production and consumption. The expansion of mass production, based on the rolling production line first introduced by Henry Ford, required the use of electric motors, but also stimulated the creation of large numbers of well-paying jobs. Mass transportation of goods internationally also stimulated the rise in the use of oil for freight transport. However, a novel feature of the fourth surge was the rise of energy-intensive forms of consumption, both in the home and for leisure travel.

With around 8 million cars being produced annually, car ownership doubled in the US in the 1950s to 67 million vehicles by 1958. This was facilitated by state investment in highways and bridges that produced a high quality road system. In turn, this encouraged householders to move out of city centres into the newly expanding suburbs and to commute to and from work in the city. Land developers were able to buy land cheaply just outside city limits in order to build cheap mass housing, with nearly 1½ million new houses being built annually in the decade after the war. Local councils competed to attract development, for example, by building schools and introducing zoning laws. The flipside of this expansion was the decline in population and investment in inner cities, as the tax base collapsed, and the elimination of mass electric streetcar transit systems in most cities.

By the end of the 1950s, almost all American households had electric lighting and refrigeration, and nearly three-quarters owned a washing machine and a vacuum cleaner. This was enabled by the spread of the supply of electricity, most of which was generated by the burning of coal. These and other electrical appliances significantly reduced the time and drudgery of housework, most of which was undertaken by women. Along with changing social attitudes, this also helped to enable higher levels of participation in paid work by women, which provided a further stimulus to economic growth. Thus, mass consumption was associated with a significant increase in the quality of life for most women and men.

The role of the state

The state played significant roles in the deployment of the mass production and consumption paradigm. In addition to local highway development, the building of the Interstate Highway System was facilitated by President Eisenhower in the Federal Aid Highway Act of 1956. The 48,000 miles of interstate highway took 35 years to complete, stimulating investment of $114 billion, with 70 per cent of the funding coming from the Federal government, mainly through fuel taxes.

A significant stimulus to conversion of the war effort to civilian purposes was provided by the so-called G.I. Bill of 1944, which provided World War II veterans with cash for technical school or college training, low-cost mortgages and low-cost loans for those wishing to start a business. By 1956, nearly 8 million veterans had made use of this support to attend college or university or receive some form of training. This significant expansion in human capital is thought to have contributed to the high levels of economic growth that the US experienced after the war.

With infrastructure investment and a highly trained workforce, this enabled private sector investment in job-creating production, both in manufacturing industries and in the rapidly expanding service sectors, such as retailing, mass entertainment and mass tourism.

In addition, in the light of the Cold War between the US and the Soviet Union until the fall of the Berlin Wall in 1991, high levels of government spending continued in research and development of new military equipment and weapons systems in what Eisenhower called the military–industrial complex.[1]

Spread of deployment to other industrialised countries

Other industrialised countries in Europe and Japan gradually emerged from post-war austerity, partly thanks to aid for rebuilding from the US under the Marshall Plan. The full deployment of the mass production and consumption paradigm was then possible, as the technologies and industries developed in the US spread to these countries. This involved overseas investment by American companies, further innovations made in these countries and adoption of mass production management methods by European and Japanese firms. This enabled significant improvements in labour productivity, resulting in average annual rates of GDP growth of 2.5 per cent in the UK, over 3 per cent in the US, 6 per cent in Germany and over 9 per cent in Japan.

The growth of trade unions enabling negotiations for wage increases to workers that reflected labour productivity improvements ensured that this wealth was equitably spread throughout these economies. State investment in unemployment benefits and national health and social care systems, particularly in European countries, also helped to ensure that the economic gains were widely spread, and to create a positive environment for private investment.

International economic management

The basis for the international economic and monetary system following the Second World War was laid down in the Bretton Woods agreement by the Allied powers in 1944. This set up a system of fixed exchange rates between participating nations, whereby national governments would act to peg their currency to within 1 per cent of a fixed exchange rate with the US dollar, which acted as a global reserve currency. In turn, the US government would link the dollar to the value of gold at a fixed rate of $35 per ounce of gold. This created a stable international monetary system that promoted international trade, further stimulating economic growth.

The Bretton Woods agreement also created two new international institutions: the International Monetary Fund (IMF) and the World Bank. The role of the IMF is to police the system and to lend reserve currency to countries faced with balance of payments deficits, i.e. when the value of their exports was much lower than the value of goods and services imported, as happened to the UK in 1976. The World Bank is a multilateral development bank, which lends money to developing countries to promote economic growth.

However, a third institution – an International Currency Union (ICU), proposed by leading British economist John Maynard Keynes at Bretton Woods – failed to get agreement from the US delegation, led by Harvard PhD economist Harry Dexter White. Keynes recognised that a fixed exchange rate system, like that created under Bretton Woods, was vulnerable to instability in the event of countries having excess trade surpluses as well as trade deficits. Countries such as the US have internal surplus recycling mechanisms, such as redistribution of taxes raised in surplus regions to benefits and infrastructure spending in deficit regions, to ease tensions between richer and poorer regions. Without similar surplus recycling mechanisms on a global scale, Keynes was concerned that the Bretton Woods system would eventually become unstable.[2] For over 20 years, the US as the world's leading economy was able to maintain international stability, despite having a large trade surplus with the rest of the world, as it recycled this surplus to the rest of the world, through aid, such as the Marshall Plan, or direct capital investment in manufacturing plants and infrastructure by US firms overseas. However, as Keynes warned, the system eventually became impossible to maintain in 1971.

End of deployment phase

The early 1970s saw two significant international events that marked the end of the deployment phase of the mass production and consumption paradigm – the collapse of the Bretton Woods system in 1971 and the oil price shock in 1973. Fortunately, as we shall discuss below, 1971 also saw the birth of a new paradigm, based on ICT. As we have seen, the mass production paradigm was led by the automobile industry, together with manufacturing in a range of other industries, such as electrical products, chemicals and plastics derived from oil. The core input for this paradigm was oil, which remained cheap throughout the post-war period as massive new oilfields in the Middle East were discovered and exploited. This was underpinned by a stable international monetary system, based on fixed exchange rates, which promoted international trade.

In the 1960s, with the Japanese and German economies prospering (thanks partly to US support and development bank investment) and its own domestic consumption levels still rising, the US had gone from a position of trade surplus to a trade deficit, as the value of imports exceeded the value of exports. At the same time, the US government was running a budget deficit, as its income from taxation was insufficient to fund high levels of government spending, including that on the Vietnam War and President Lyndon Johnson's Great Society programme of public spending to combat poverty. As the US was the world's reserve currency, it could finance this budget deficit by continuing to print dollars to buy foreign imports. However, this led to monetary inflation, weakening the value of the dollar and making it hard to maintain the peg to gold at $35 per ounce. In August 1971, President Richard Nixon unilaterally ended the convertibility of dollars into gold, effectively ending the Bretton Woods system of fixed exchange rates. Within two years, all major countries had moved to a system of floating exchange rates between their currencies. At the same time, the Nixon and subsequent US administrations began to squeeze labour costs by limiting union power in wage bargaining in order to improve the competitiveness of US companies, and to allow the Federal Reserve to increase US interest rates to combat inflation.

Oil price shocks

To meet the growing demand for oil in Western countries in the fourth surge, global crude oil production grew rapidly from 10 million barrels per day in 1950 to 55 million barrels per day in 1973.[3] North America (the US and Canada) remained the largest producing region until 1966, when it was overtaken by the Middle East (principally Saudi Arabia, Kuwait, Iran and Iraq) after a number of super oilfields were discovered there. However, during this period, the global oil price remained remarkably stable at around $3 per barrel. This was partly owing to the efforts of the Organisation of Petroleum Exporting Countries (OPEC). OPEC was formed in 1960 by the Middle Eastern oil exporters and Venezuela as a cartel to coordinate supply and maintain oil prices, resisting efforts of multinational oil companies (known as the 'Seven Sisters', mainly based in the US) to force price reductions.

Following US support for Israel in the 1973 Arab–Israeli War, in October 1973, the Arab members of OPEC (Saudi Arabia, Kuwait, Iraq, Libya, United Arab Emirates and Qatar) plus Egypt and Syria began an embargo on oil exports by cutting production by 5 per cent every month and blocked all oil deliveries to the US. They also imposed increases in the price of their oil exports, causing the international market price of oil to jump from $3 per barrel to $12 per barrel by the end of 1974. Though the oil embargo was lifted in March 1974, the continuing high oil prices had major effects on consumption and national economies. The immediate economic impact was a rise in gasoline prices for US consumers from 38 cents per gallon in May 1973 to 55 cents per gallon in June 1974, causing long lines at gas stations in some States. The US and other Western countries, including the UK, suffered an economic recession from 1973 to 1975, with high levels of unemployment, falling or stagnant economic growth and high price inflation, known as stagflation.

With global oil supply only rising slowly and continuing high oil demand despite the economic downturn, oil prices remained high during the 1970s. This caused a huge surge in income for the oil producing countries in the Middle East, creating further ripples in the global economy, as discussed below. In 1979, the revolution in Iran and subsequent reduction in their oil production caused a second oil shock with global oil prices doubling to a then-high level of $39.50 per barrel in 1980. This again led to panic buying and long lines at gas stations in the US.

The role of oil in the economy

Given the role of oil in both production and consumption in the fourth surge, it is clear that the low global price of oil from 1950 to 1973 was a key driver of the deployment of this paradigm, and that the high oil prices of the later 1970s marked the end of this deployment phase. We have focused on the role of the US as it was the dominant international producer of goods and services at the start of this phase and evolved into the dominant international consumer by the end of this phase. Three interlinked factors helped to drive this transition: the collapse of the Bretton Woods system in 1971, the rise in global oil prices and the peaking of US domestic oil production in 1970. However, the causal relations between these factors are not easy to disentangle, as conventional macroeconomic models of economic growth do not include energy as a key input, as noted in Chapter 3.

Oil is a finite resource, so concerns have long been raised as to limits to its supply. The most famous of these concerns was voiced by US geophysicist M. King Hubbert at a meeting of the American Petroleum Institute in 1956.[4] On the basis of information about past discoveries, Hubbert projected that US domestic oil production would peak between 1965 and 1970. In fact, US oil production peaked in 1970 at 9.6 million barrels per day, and slowly declined after that until 2008 at a rate similar to that which Hubbert projected. Since then, a number of experts, mainly geophysicists, have tried to apply similar ideas of 'peak oil' at a global scale, whilst other experts, mainly economists, argue that there are currently many years of known oil reserves, which can be economically extracted if there is sufficient demand. As discussed in Chapter 3, this can be thought of as a race between depletion of resources and technological improvements in the finding and efficient extraction of those resources. In the US, Hubbert's projection has broken down in recent years with new extraction techniques in the form of hydraulic fracturing (fracking) enabling economic extraction of shale oil reserves, leading to a significant increase in US oil production since 2008.

From a national economy perspective, the key factor relating to oil is the price of oil as an input into the economy. The low price of oil was a key driver of national economic growth in the US and other Western countries from 1950 to 1973. However, the stable global macroeconomic environment provided by the Bretton Woods system was also clearly an important factor. This gave a stimulus to international trade, whilst assigning a strong role to national governments in managing economies so as to promote economic growth, along the lines recommended by Keynes. The near simultaneous collapse of this international system with the global oil price shock and the peaking of US oil production in the early 1970s could have been disastrous for the US and thus the global economy, but for three new factors: a new informal international financial system, a new neoliberal economic orthodoxy and the installation phase of the fifth techno-economic paradigm based on ICT. These factors helped to stimulate global economic growth from the mid-1980s until the economic crash of 2008, but the first two arguably helped to lay the basis for that economic crisis, the repercussions of which are still being felt as we move into an uncertain future in 2017.

A new international financial system

Every year since 1976, the US has run a trade deficit, i.e. the value of goods and services imported were higher than the value of exports, so that it is consuming more than it is selling to the rest of the world. At the same time, for every year since 1970, except for 1998 to 2001, the US government has had a budget deficit, i.e. the value of government spending was higher than that of tax receipts. Economists would normally expect it to be unsustainable for a country to run these twin deficits for such a long period, except for the advantages that the US uniquely possesses.[5] First, the fact that the US dollar is still the main reserve currency that the rest of the world uses, including for pricing sales of oil and other commodities. This means that the dollars earned by oil exporting countries, which due to high oil prices were much greater than they could use to finance their own economies, were recycled to the US by buying US Treasury bonds (basically, IOUs from the US government) or investing in shares in US companies or Wall Street financial firms, such as Goldman Sachs. Similarly, surpluses generated by exporters in Japan and Germany, and more recently in China, were recycled to the US. Second, the fact that the US still had a substantial domestic oil production, rather than having to import all its oil needs, meant that higher global oil prices improved the competitiveness of US firms compared to its competitors in Japan and Germany. Third, despite an improvement in labour productivity in the US, this was not reflected in a rise in real average US wages. Fourth, US geopolitical power reflects continuing high levels of military spending.

A new neoliberal economic orthodoxy

With the election of Prime Minister Margaret Thatcher in the UK in 1979 and President Ronald Reagan in the US in 1980, a new strand of economic thinking gained dominance. This is generally referred to as neoliberal economics, drawing heavily on the thinking of US economist Milton Friedman and Austrian economist Friedrich Hayek, whilst relegating the previously dominant ideas of John Maynard Keynes to history. Neoliberal economics emphasises the role of free markets and free trade in promoting economic growth, and generally argues for governments to 'get out of the way' and let markets rip. US and UK governments thus reduced government spending (except on the military), introduced tax cuts to stimulate business activity and private consumption, and deregulated the financial system, e.g. by repealing the Glass–Steagall Act that separated high-street from investment banking. Despite leading to a growing budget deficit in the US, this strategy appeared to be working in the late 1980s and 1990s, as high levels of consumption, increasingly funded by private debt on credit cards and short-term loans, house price inflation and ever more complex financial instruments stimulated economic growth. Even though this led directly to the financial and economic crisis of 2008, neoliberal economic thinking still seems to be alive and well in mainstream academic economics departments and think tanks influential on government policies, such as the Cato Institute and the Heritage Foundation in the US.

Installation phase of the fifth surge

The long-run historical perspective being followed in this book attributes a greater role in the global economic growth of the 1980s and 1990s to the installation phase of the fifth major economic surge, based on information and communication technologies. As with the other paradigm changes discussed, the ICT surge had a long gestation period even prior to its installation phase, but it is convenient to identify the beginning of its installation with the development of the first commercial microprocessor, i.e. a general purpose computer on an integrated circuit, by Intel in 1971. As with earlier technological breakthroughs, the commercial potential of the microprocessor was not immediately apparent, even to the company developing it. At the time, computers were generally large mainframe machines used mainly for numerical calculations. Later versions of the Intel microprocessor were made available to computer hobbyists, such as Bill Gates and Paul Allen who then developed the Microsoft operating system that became the standard for personal computers, and Steve Wozniak and Steve Jobs who developed the first Apple computer in 1976. These developments were enabled by an exponential growth in computing performance, as the number of transistors on each integrated circuit doubled every two years. This rate of improvement is known as Moore's law, after Gordon Moore, co-founder of Intel, which became a guide to future performance for the computing industry.

The first programmable computers were built during the Second World War for code breaking and other military applications, led by pioneers Alan Turing and Tommy Flowers in the UK and John von Neumann in the US. These computers used vacuum tubes or thermionic valves to control the flow of electricity in order to undertake calculations, but these were energy intensive and prone to over-heating. Later computers were controlled by semiconductor transistors, which were developed at Bell Labs in 1948 by John Bardeen, Walter Brattain and William Shockley. Bell Labs was a leading research and development facility that had been set up in 1925 by the American Telephone and Telegraph Company (AT&T), as part of a deal with the US government to allow it to have a monopoly on the provision of telephone services in the US. Bell Labs went on to develop many revolutionary technological breakthroughs, leading to eight Nobel Prizes, including one for the transistor.

The other necessary element for the ICT surge was the development of the Internet for communication between computers. This was originally developed by the Advanced Research Projects Agency (ARPA) of the US Department of Defense in the late 1960s and for civilian networking under National Science Foundation grants in the early 1980s. The next crucial development was the World Wide Web in 1990 by British scientist Tim Berners-Lee, working at the CERN nuclear research centre in Switzerland, which used hypertext to link and access information between computers over the Internet. These and subsequent developments enabled the power of computing to diffuse widely both in businesses and for personal use, with word processing and spreadsheet applications becoming ubiquitous in office settings and for management of supply chains.

Dot-com bubble

As in previous surges, the possibilities opened up by the new technologies led to a wave of financial investment in the new firms which had sprung up to take advantage. This investment wave was further enabled by the deregulation of financial markets associated with the rise of the neoliberal economic paradigm. This created what became known as the 'dot-com' bubble in the late 1990s, named after the website addresses of firms operating mainly or wholly on the Internet. Many of these firms attracted high valuations when they launched on to stock markets through initial public offerings (IPOs) due to expectations of future earnings, despite making little or negative profits at that time. Inevitably, the bubble burst in early 2000, as it became clear that the high valuations of many of these dot-com firms were unrealistic. The NASDAQ stock exchange in New York, on which most of these firms were traded, peaked in March 2000 and lost half of its value by the end of that year. As a result, many dot-com firms went out of business, causing huge losses for investors. In succession, this downturn affected the rest of the economy and, compounded by uncertainties created by the 9/11 terrorist attacks in New York and Washington, DC, US stock markets continued to decline, losing $5 trillion in market value from March 2000 to October 2002.

To some extent, this led to a shake-out with only those Internet companies with robust business models surviving. The most successful surviving companies went on to have huge market valuations by dominating in their respective markets, including Google in Internet search, Amazon in online sales and Facebook in social networking. However, concerns have been raised that these business models rely on harvesting data from users in order to direct relevant advertising towards those users, raising fears over privacy and the level of social control that these firms are able to enact.[6]

The next step in mass computing use came via the melding of mobile communications and computing power in smart phones. The most successful of these was Apple's iPhone, launched in 2007, which helped Apple to become the company with the highest market valuation in the world by 2012. Though this success is often attributed to the company's design process and the salesmanship of its founder Steve Jobs, most of the features that make the iPhone so attractive to users, including touchscreen, synthetic voice, GPS positioning and Internet and web access, were originally largely publicly funded, as economist Mariana Mazzucato has pointed out.[7] Further concerns have been raised about the low levels of tax being paid by computing and online companies, as they organise their business arrangements to reduce taxable profits in jurisdictions such as the US and the UK, which have higher levels of corporate taxation.

Financial crisis of 2008 and its aftermath

If history was just on a repeating cycle, then it would suggest that the shake-out following the crash of the dot-com bubble in the early 2000s would have given way to a golden age of prosperity linked to the deployment of the ICT paradigm. For a while, it looked like this might be possible, at least to central bankers in the US and the UK. In a speech in 2004, US Federal Reserve Governor Ben Bernanke claimed that we were living in a time of 'the Great Moderation', with a steadily growing economy in which major fluctuations in growth rates had been smoothed out, largely thanks to the inflation targeting being undertaken by independent central banks.[8] Similarly, in 2003, Bank of England Governor Mervyn King said that the UK was experiencing a 'nice' decade, standing for 'non-inflationary consistently expansionary' period of steady growth, falling unemployment and growing levels of wages and consumption, with little or no price inflation.[9]

Sadly, it turned out that the structural imbalances in the economy were much greater than most economists or central bankers realised. A broader awareness of historical patterns might have helped. We now know that there was over-investment in assets in the real economy, particularly lending for housing in the US to people who would struggle to pay back the loans – the so-called subprime mortgages. Instead of managing these risks, banks and financial institutions packaged these subprime mortgages into complex financial securities, such as collateralised debt obligations (CDOs), which were then sold on as triple-A rated safe assets. When house prices started to fall in 2006 and higher levels of interest started to kick in, many subprime mortgage holders began to default on their loans. The effects of this began to spread throughout the economy, as banks gradually realised that they had no idea of the value of mortgage-backed securities on their or other banks' books, so they became increasingly reluctant to lend to each other. The monetary merry-go-round stopped on 15 September 2008, when US investment bank Lehman Brothers declared bankruptcy, after US Treasury Secretary Hank Paulson had refused to authorise a bail-out. Facing the potential for bankruptcy of further big financial firms, such as Merrill Lynch and AIG, the US government initiated bail-outs of these firms and launched a $700 billion Troubled Asset Relief Program (TARP) to temporarily buy up these toxic assets. This and similar measures in other industrialised countries helped to avert a complete breakdown of the financial system. Nevertheless, between 2008 and 2010, the US economy lost nearly 9 million jobs and GDP contracted by over 5 per cent. Similar contractions occurred in other industrialised countries, though emerging economies such as China suffered a smaller impact on their economies.

Again, it might have been hoped that this shake-out would have signalled the turning point of the fifth surge, with an expansionary deployment phase following this point. However, the economy has not returned to a normal state. Interest rates in the US and other industrialised economies remained at ultra-low levels by the start of 2017. The jobs lost in the recession were recovered by 2014, though it is estimated that the majority of new jobs were low-paying jobs. Part of the recovery can be attributed to stimulus of around $800 billion provided by the American Recovery and Reinvestment Act of 2009 in the form of increased government spending on health care, education and infrastructure and tax credits and other social welfare benefits, which helped to stimulate economic activity. However, this also contributed to an increase in the US national debt, which currently stands at just under $20 trillion, with the US Treasury having to apply 'extraordinary measures' to maintain this debt ceiling, which was reimposed in March 2017. In addition, US household debt has risen again to over $12 trillion.

Next steps for the global economy

The start of 2017 marks a time of great uncertainty for the global economy. The historical record suggests that there is scope for a deployment phase of ICT technologies that could drive a new phase of economic growth. However, past surges were driven as investment switched from the bubble period dominated by financial capital to a period of investment in the real economy led by production capital. This investment also gradually spread out from the leading country to other industrialised countries. Three factors complicate the potential for this deployment phase to take hold. First, the majority of the new investment needs to be in industries and services that dramatically reduce global carbon emissions, in order to meet emissions reduction targets agreed at the UN Climate Change Conference in Paris in 2015. Though Carlota Perez and Mariana Mazzucato have argued that there are clear synergies between the features of ICT technologies and low carbon technologies, these will require significant institutional changes and adoption of new business models, such as circular economy models.[10] Second, industrialised economies have become much more financialised, in that value creation relies much more on the manipulation of financial assets rather than productive action, even by firms that are thought of as in the productive economy. Third, as we have seen, the global economy has become much more complex with the US retaining a dominant position, despite twin budget and trade deficits, whilst China has undergone a rapid economic expansion, which it is now struggling to maintain. We shall return to discuss the implications of these issues, after we have looked more closely at the theoretical understanding of the relations between energy use and economic growth in the next chapter.

Notes

1 Eisenhower (1961).
2 Varoufakis (2015).
3 See BP Statistical Review of World Energy (no date).
4 Hubbert (1956).
5 Varoufakis (2015) refers to these features as the Global Minotaur.
6 Morozov (2014).
7 Mazzucato (2013).
8 Bernanke (2004).
9 King (2003).
10 Perez (2013); Mazzucato and Perez (2014); Perez (2016).

PART III

Implications for economic development

8

ENERGY AND ECONOMIC GROWTH – A POSITIVE FEEDBACK SYSTEM

Introduction

We have now seen that technologies associated with the increasing use of energy played a key role in the five historical surges of economic growth, from the first industrial revolution to the post-Second World War boom of mass production and consumption. As the history of these surges shows, though, the deployment of these technologies coevolved with a range of social and institutional factors, including changes in consumer practices and advances in the roles of regulatory and financial institutions. We now want to draw out the implications of these ideas for understanding the macroeconomic effects of a low carbon energy transition. So, what we really need is a theory of economic growth that incorporates the interactions between the energy dependence of economic activities and the wider coevolutionary developments in the economy. Unfortunately, such a theory does not yet exist. Mainstream understanding of economic growth has severe limitations from our perspective and does not include a role for energy inputs.

However, we are able to draw on several interesting lines of thinking that illuminate different aspects of the relations between energy and economic growth. In particular, we want to emphasise ideas on the role of energy in the macroeconomy that seem to be consistent with our coevolutionary view of economic development. As we saw in Chapter 3, this view seeks to combine an understanding of economic change as involving dynamic interactions between technologies, institutions, business strategies and user practices with dependence on ecological systems. The role of energy inputs is a key part of this ecological dependence. In this chapter, we examine recent work showing that the availability and efficient conversion of energy sources to useful work have been key drivers of economic growth. In particular, these types of improvements associated with electrification and the rise of oil for transportation have been key in the last two major surges.

We argue that there are systemic positive feedbacks between the availability of useful energy and growth in economic activity, so that each supports and reinforces the other. A better understanding of these systemic feedbacks is important for appreciating the impacts of a transition to low carbon energy sources on economic growth going forward.

Productivity improvements

In simple terms, economists explain the long-run engine of growth in an economy in terms of increasing productivity. This means getting more output out of the inputs going into economic activity. What are the key inputs? First, labour, which is usually measured in terms of the number of hours worked in economically productive activity. Second, capital, which is a measure of the buildings, machinery and equipment used in production processes. The focus is on labour productivity, i.e. getting more economic output out of each hour worked. If labour productivity can be increased, then the total value added in the economy will grow. Provided that some of this additional value goes back to the workers, e.g. in the form of increasing wages, then they will be able to buy a proportion of the additional goods and services in the economy, so increasing their wellbeing (other things being equal). One way of increasing labour productivity is by people working harder or smarter. For example, Adam Smith's division of labour argument shows that a group of workers can produce more if each one concentrates on the part of the process where they are most skilled. Accumulating capital by investing in machinery and equipment can also enable workers to be more productive. However, it was found that just expansion in the amount or productivity of labour and capital were not enough to explain economic growth. An additional factor relating to technological change needs to be included. This raises the overall productivity of any particular combination of labour and capital.

Modelling economic growth

The dominant mainstream model of economic growth was formulated by US economist Robert Solow in 1956.[1] In this model, the output of an economy, measured by its GDP, is a function of capital and labour inputs to the economy, together with a factor A which does not depend on these inputs and represents any shift in production possibilities, including technological or institutional change. On one level, this model was very successful in that it fitted very well the growth of US GDP from 1909 to 1949. Over this period, output per hour worked doubled. However, in order to do this, nearly 90 per cent of this growth was attributed to the factor A, called the residual, which only has a general explanation as technical change. This factor is now referred to as total factor productivity (TFP), but the model does not provide a good understanding of what drives this, so it has been referred to as a 'measure of our ignorance' of the drivers of economic growth.[2]

Though the Solow model is conceptually incomplete, it provides a clear theoretical model for understanding economic growth. From our perspective, though, in addition to a lack of explanation of technological change, the Solow model is limited in other ways. It describes a process of growth in equilibrium, so that, whilst the economy as a whole is growing, total supply and demand in the economy match at any point. The model also assumes that different factors of production, such as labour and capital, are substitutable for each other, e.g. that there are no limits to the replacement of labour by capital in an economic process, or vice versa. This is important because, in as far as energy use is included in the model, it is through the use of capital such as machines. This principle implies that there is no minimum to the amount of energy used for any productive process, which is unrealistic.

In the 1980s, work by other US economists, including Paul Romer and Robert Lucas, developed a revised theory of economic growth, in which growth is explained by factors within the theory, including knowledge and human capital.[3] Human capital relates to the level of skills and abilities of the workers within the economy. This so-called 'endogenous growth theory' has formed the basis for mainstream economic and policy analysis, being quoted by politicians such as Gordon Brown, the future British Chancellor of the Exchequer and prime minister, in a speech in 1994.[4] Clearly, as our historical studies show, the role of human skills and knowledge is crucial to the development and spread of new growth surges. However, from our perspective, this theory is still limited in that it describes a process of growth in equilibrium and does not incorporate energy or natural resource inputs into the economy.

Energy–economy growth engine

An alternative approach has been pioneered by ecological economists Robert Ayres and Benjamin Warr,[5] building on the work of Nicholas Georgescu-Roegen discussed in Chapter 3. Ayres and Warr describe the role of energy as a driver of economic growth through a series of positive feedback loops. Cheaper energy inputs to economic activities result from new discoveries of energy sources, economies of scale and technological progress through learning effects in the conversion of primary energy sources to useful energy inputs. These cheaper energy inputs reduce the cost of goods and services produced, stimulating the demand for those goods and services, leading to an increase in economic output. As a large proportion of the value of that increase in energy output goes to increased wages (and only a small proportion goes to returns to energy producers), this increase in wages will stimulate further effort through R&D and innovation to substitute energy-using machines for labour. This will increase energy inputs and stimulate further scale and learning improvements. This feedback process can continue as long as further cost reductions in a type of energy technology or process for meeting particular end-use demands can be found. For example, the demand for electricity stimulated significant scale and technological improvements in power stations for burning coal to generate electricity. In addition, as we discuss further

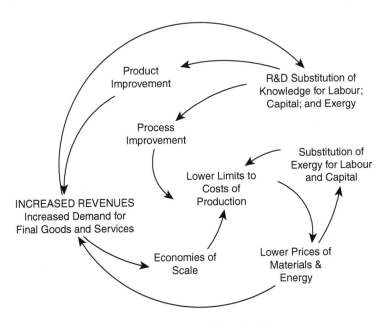

FIGURE 8.1 The Ayres–Warr endogenous growth mechanism.
Source: Warr and Ayres (2012), p. 96.

below, this feedback process can lead to spillovers to other end uses. The feedback loops are represented in Figure 8.1.

So, why is energy not considered a significant factor of production in mainstream economic growth models? Warr and Ayres point to three main reasons.[6] First, as with other aspects of the natural environment, energy sources are not assigned any intrinsic value in mainstream economics and so the costs of extracting energy are included in the costs of machinery and equipment required to extract them, such as coal mines or oil drilling, which is considered as part of capital spending. Second, these growth models generally assume that different factors of production are independent, whereas in fact, labour, capital and the services provided by energy contribute together. For example, harvesting a field requires human labour, capital in the form of scythes or combine harvesters and energy in the form of horse power or mechanical power for the harvesters. Third, these models generally only consider energy inputs into the economy, such as coal, oil or gas, as intermediate factors of production. Warr and Ayres argue that what matters in terms of the economy is the useful work that these energy inputs provide. As we discussed in Chapter 2, in any energy conversion process, the total energy is always conserved but some energy is always lost in the form of waste heat. The useful part of energy, in terms of its ability to do work, such as power a machine, is measured in terms of a quantity called exergy. The useful work is thus measured by the exergy delivered into the economy, for example the power that moves a machine or vehicle forward or the light that illuminates a room.

Modelling the role of energy in economic growth

This picture of substitution of exergy-using machines for labour driving a reduction in the cost of goods and services provides an intuitive picture of the role of energy in economic growth. However, to provide a more quantitative analysis, most mainstream and alternative economists use a production function approach. This specifies how economic output is a function of the inputs into the economy. As we discussed previously, mainstream economic modelling takes labour and capital as the key inputs, or factors of production. Earlier classical economists, such as Adam Smith, who were analysing a predominantly agricultural economy, took land as a third factor of production, but this was argued to be less relevant by later economists analysing industrial economies.

An ecological economic perspective would take some measure of energy inputs as a third factor of production. Though some mainstream economists looked at introducing energy and materials into production functions in the 1970s, the influence of energy on determining outputs was argued to be minimal. This is because of a fundamental assumption, known as the 'cost-share theorem'. In the US economy, energy accounts for only around 5 per cent of the input costs, with capital accounting for 25 per cent and labour accounting for 70 per cent.[7] In a simple market economy in equilibrium, these cost shares would determine how much output changes in response to a change in the weight of that input in the total output. For example, a 1 per cent reduction in energy consumption, with no change in labour and capital, would then reduce output by 0.05 per cent. In 1985, German theoretical physicist Reiner Kümmel and colleagues showed that if energy is included in a production function and the economy has multiple productive sectors and is not assumed to be in equilibrium, then the cost-share theorem no longer holds.[8] The upshot of this is that the economic importance of energy inputs is much greater than their relatively small share of total costs. This is the mathematical version of the physical argument for the importance of energy in economy growth.

Kümmel and colleagues then derived a new model of national economies in which economic output is a particular function of labour, capital and exergy inputs into the economy. They showed that this model reproduced the real growth in output for the German and US economies for the period 1960 to 1981 or 1978 (though this still required an unexplained factor increasing over time). In his later work, Kümmel added human creativity (closely related to the concept of human capital) as a fourth input into economic growth to explain the time trend, and reproduced German and US output up to 2000. From this work, Kümmel formulated a 'second law of economics' stating that 'Energy conversion and entropy production determine the growth of wealth', which he expounded in a detailed book.[9] Whilst we agree that economic growth is dependent on the efficient conversion of energy inputs, it may be going too far to imply that this is the determining factor. As we have seen, knowledge and social institutions are also important in determining the rate and distribution of economic growth.

Linking ecological and evolutionary economic perspectives

In parallel to the work of Kümmel, American-born physicist and economist Robert Ayres was developing and testing a physical-based approach to understanding economics. In 1969, with fellow American environmental economist Allen Kneese, he pioneered a view of economies as harnessing energy flows and processing materials to provide useful services, which naturally leads to the creation of waste heat and physical wastes.[10] Later, as already discussed, working with British ecological economist Benjamin Warr, he set out to see if this physical-based approach could account for the unexplained residual in Solow's model of economic growth. In the book that outlines the evolution of their thinking, Ayres and Warr describe how they take a 'disequilibrium (quasi-evolutionary) approach [that] characterizes the economy at the macro-level as an open multi-sector materials/energy processing system'.[11] Drawing on evolutionary economist Joseph Schumpeter, they argue that long-term economic growth has been driven by radical technological breakthroughs ('creative destruction') associated with improvements in the efficient conversion of exergy inputs into useful work and services that people demand. This picture is very much in line with the coevolutionary view of techno-economic change that we have expounded here (indeed, their book was one of the inspirations for the present author). However, they did not link their work explicitly with Freeman and Perez's detailed historical and theoretical analysis of techno-economic surges, as we have done here.

Useful work as an input

In order to go from a general argument to be able to undertake a quantitative analysis, Ayres and Warr also followed a production function approach, though they noted that there are theoretical problems in aggregating a range of different types of economic activity into a simple production function.[12] In fact, the quantitative production function model that they develop only partially incorporates the positive feedbacks in their intuitive energy–economic engine of growth shown in Figure 8.1. This is because a full theoretical model of the economy requires a theory of how the value of the outputs from the economy is distributed back to workers and owners of capital and energy-using machines. If most of the value goes back to workers in the form of wages, then this enables those workers to buy the goods and services produced by the economy.[13]

They used the same production function as Kümmel and colleagues, but instead of exergy, they used useful work. As we saw in Chapter 2, useful work is the exergy into the economy (from primary energy sources of fossil fuels, nuclear or renewables) multiplied by the efficiency of conversion to the final stage at which work is done. There are four main categories of useful work: (1) mechanical drive to drive machines and vehicles; (2) high- and low-temperature heat for industrial processes such as steel production; (3) electricity; and (4) muscle work by people and animals. Ayres and Warr argued that useful work is the closest thing to what people want from the economy that can still be measured in energy units, and so can be included in this type of aggregate analysis. In the next section, we discuss the final stage of conversion of useful work to provide the services that people actually want, such as mobility.

Ayres and Warr argued that when they introduced a function of useful work alongside labour and capital into their production function, they could reproduce the time series of US GDP from 1900 to 2000, without including Solow's residual function representing technical change or any time-dependent parameter (see Figure 8.2). This implies that increases in primary energy and improvements in the efficiency of its conversion to useful work, acting in synergy with other labour and capital inputs, can explain the otherwise unexplained contribution of technological change to economic growth. However, as Ayres and Warr used an unconventional production function in their analysis, this insight has been ignored by mainstream economists, and so further research is needed to better understand their findings.

In their more recent work, Warr and Ayres contend that, to reproduce US GDP growth more accurately after around 1975, they need to include an additional factor of production relating to ICT.[14] They argue that ICT is enabling new opportunities for creating and capturing economic value, similar to past general purpose technologies (GPTs) such as the steam engine and electrification (whose properties we discuss in the next chapter). These include new ways of improving the efficiency of conversion of primary energy to useful work, such as smart grids that enable two-way flows of electricity across networks.

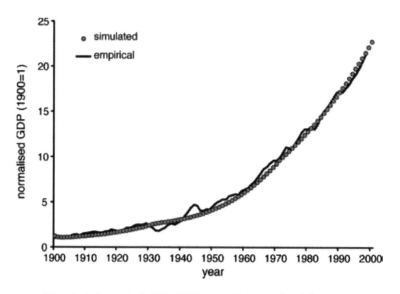

FIGURE 8.2 Historical (empirical) US GDP and GDP simulated by Warr and Ayres' model, 1900–2000.

Source: Warr and Ayres (2006), p. 346.

Understanding efficiency of conversion to useful work

In research led by PhD student (now Dr) Paul Brockway with Julia Steinberger, John Barrett and myself, we examined how the overall efficiency of energy conversion in the UK, the US and China has evolved, and the implications of this for projections of future energy demand and carbon emissions.[15] Building on the work of Ayres and Warr, we examined the efficiency of conversion of primary exergy into these economies into useful work done in the economy for the period 1960–2010 (or 1971–2010 for China, due to limitations on the earlier data). As we discussed in Chapter 2, exergy efficiency is defined as the ratio of useful work out to primary exergy in, and so useful work is just the product of primary exergy input (which is more or less the same as primary energy input) and exergy efficiency:

$$\text{Useful work out} = \text{primary energy in} \times \text{exergy efficiency}$$

This research produced several interesting findings, presented in Figure 8.3. First, for all these economies, at most 15 per cent of the primary exergy, i.e. the energy available to do work mainly in the form of coal, oil and gas, is converted into useful work done. This is because of the high level of losses at each stage of energy conversion.

Second, there seems to be a tendency for improvements in exergy efficiency to slow down over time. Whilst in the UK, the overall exergy efficiency of the economy has risen from 9 per cent in 1960 to 15 per cent in 2010, in the US, the overall exergy efficiency has remained remarkably constant at around 11 per cent over this whole period. How could it be that, despite improvements in conversion efficiencies for devices for most end uses, the overall conversion efficiency has

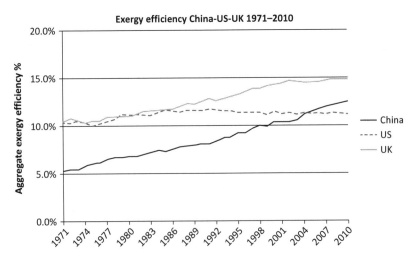

FIGURE 8.3 Exergy efficiencies for China, the US and the UK, 1971–2010. Source: Brockway, based on Brockway et al. (2014, 2015).

remained constant? What we found was that, as the US has moved from a manufacturing-based economy to one based on services and high levels of consumption, the mix of economic activity has shifted so that less efficient end uses have replaced more efficient end uses. For example, air conditioning, which has risen from 10 per cent to 20 per cent of electricity use, is highly inefficient in providing the service of keeping buildings cool in hot weather. This could be achieved much more efficiently, with lower primary energy input, by designing or refitting buildings with natural ventilation.

Third, this tendency for exergy efficiency improvements to stagnate in economies could have important implications for future end use. From 1971 to 2010, China's overall exergy efficiency showed a 2.5-fold increase from 5 per cent to 12.5 per cent, as it underwent a rapid industrialisation. This meant that a four-fold increase in primary exergy into the economy was converted into a ten-fold increase in useful work done in the economy. As China continues its rapid economic growth into the future (even at expected lower rates of growth than over the past 20 years), the useful work required for its economic activity is expected to double again by 2030. However, the structure of its economy is expected to gradually change more toward services and domestic consumption. Following the pattern of the UK and the US, this would suggest that it would not maintain its trend of exergy efficiency improvements, and these would level off at around 13 per cent. Our calculations then suggested that China would need a higher primary energy input than that projected in comparable scenarios produced by organisations such as the International Energy Agency and the energy company BP. Without continuing improvements in conversion efficiency, China could require as much as three times its current primary energy demand by 2030 to maintain its projected levels of economic activity, with significant implication for its future carbon emissions (unless it is able to significantly increase its rate of deployment of low carbon energy technologies).

In simple terms, if an economy requires an increasing level of useful work to continue to grow its economic activity, then that useful work either has to come from putting more primary energy into the economy or improving the efficiency of converting that primary energy to useful work. If there are limits to conversion efficiency improvements, then that implies a higher level of primary energy input to maintain the projected economic growth. This illustrates the importance of better understanding the energy dependence of economic development. In Chapter 11, we discuss the potential for economic activity to be further decoupled from useful work inputs – a so-called dematerialised economy; but, at least in the short term, significant amounts of useful work will be needed for agriculture, manufacturing activity, transport of goods and people, and powering electrical equipment, including information and communication technologies.

Energy services perspective

So far, we have looked at energy demand in terms of the ability of the economy to provide useful work. This is valuable for looking at trends in the whole economy, as different types of useful work can all be measured in common energy units. However, to further illuminate the relations between energy use and economic growth, we need to think about what drives the demand for this energy from the user's perspective. It is not directly the useful work that is wanted, it is the services that this provides. This difference is important because improvements in the efficiency of providing that service have been important in increasing the value to the user whilst reducing the physical amount of energy used and the associated resource needs and environmental impacts. The first energy service demand is heat for cooking and warmth in the home. As we noted in Chapter 4, billions of people in poorer developing countries still rely on biomass in the form of wood on open fires to provide these services. This is intensive in time needed for collection and is associated with high levels of indoor air pollution from wood smoke, causing breathing problems and respiratory diseases. In industrialised countries, this was alleviated by a transition to coal for heating and the technological innovation of the chimney. This moved the problem from indoors to the city level, leading to the problem of smogs, again contributory to poor air quality and respiratory diseases. This gradually precipitated a switch to cleaner burning gas and electricity for providing cooking and heating services.[16]

British energy economist Roger Fouquet has compiled and analysed several centuries of data on the costs and demand for energy services in the UK.[17] He found that, as a result of transitions in technologies from candles, whale oil, gas to electricity and improvements in conversion efficiencies, the cost of lighting services has reduced dramatically since the 1700s.[18] In 1830, using gaslights, it cost £2,700 (in current money) for one million lumen-hours (the illumination provided by a 100-Watt incandescent bulb for 30 days). It now costs just £1 to provide the same level of illumination using current highly efficient LED (light-emitting diode) lights. Not surprisingly, this reduction in the cost of lighting has led to a large increase in consumption. Each person in the UK now uses nearly 40,000 times more lighting than in 1750.[19] Similarly, transitions in technologies from horse-powered to oil-powered vehicles and improvements in conversion efficiencies for providing transport services has led to each person in the UK using 250 times more passenger kilometres than in 1750. Technological transitions from horse power and steam power to electric motors for powering machines led to each person in the UK using 60 times more power than in 1750.

Energy, technological change and economic growth

As noted, mainstream models of economic growth do not include energy or resources as inputs to the economy. This does not fit well with the importance of energy production and conversion technologies in previous surges of economic growth that we have discussed in earlier chapters. Mainstream economic growth models, such as the Solow model, built on a picture of economies as being in equilibrium. In the Solow model, the tendency of economic growth to slow as there are diminishing returns to capital and labour inputs is offset by improvements in technological knowledge which are independent of other economic activity. The picture that we have been painting shows that these technological improvements, including in energy technologies, are very much driven from within the economic system, in response to demands and opportunities, and depend on the current state of knowledge. This builds on the ideas of Schumpeter, who argued that innovation driving waves of 'creative destruction' is the key long-term driver of economic growth. So-called Schumpeterian growth models try to capture some of this by incorporating investment in research and development (R&D) into the models.[20] However, these models still do not generally incorporate energy inputs, which we have argued are crucial.

The alternative approach of ecological economics starts from a physical picture of production as a process of converting energy and resources into goods that provide useful services whilst producing waste products. We argue that a better understanding of economic growth would come from combining this ecological economics perspective with the Schumpeterian view of innovation and surges driving economic growth. This is not to say that increases in energy use alone explain economic growth. As we have seen, institutions, incentives and the state of knowledge are also crucially important in enabling the value of economic outputs to grow.

In an important step towards bringing together the mainstream and ecological economics views, Australian energy and environmental economist David Stern has developed a new model which enhances the Solow growth model by including energy as an input.[21] His model represents two regimes of economic growth, depending on energy availability. When energy inputs are scarce, economic growth depends on the level of energy inputs and the level of conversion technologies and so is constrained by their availability. When energy inputs become abundant, economic growth becomes more dependent on capital and labour inputs with technological change improving rates of energy use becoming more independent, as in the original Solow model. He argues that the former regime represents the case of Western countries before the industrial revolution and many current developing countries, in which the availability of scarce renewable energy inputs, particularly wood, limits the potential for economic growth. On the other hand, the latter regime represents the case of Western countries after the industrial revolution, in which technological change means that fossil fuel energy inputs become abundant and so economic growth is not constrained by energy inputs. Stern with Swedish environmental historian Astrid Kander subsequently showed that this model could explain well the role of the transition from traditional to modern energy sources in economic growth in Sweden from 1850 to 1950.[22]

This suggests that the Stern and Kander model captures something important about how modern economies have loosened the grip of energy on economic growth, but without breaking the grip entirely, especially if the availability of modern fossil fuels becomes limited by design or constraint. However, this model is still essentially a growth-in-equilibrium model, and doesn't fully capture the Schumpeterian evolutionary dynamics of innovation and surges. As yet, there are no fully developed economic models that combine an evolutionary dynamics with an ecological view of the energy dependence of economies.

Implications for a low carbon energy transformation

This chapter has, unavoidably, been rather dense in terms of ideas presented. It has aimed to summarise a range of different strands of analysis that provide bridges between the evolutionary economics view and the ecological economics view of drivers of economic growth. The evolutionary economics view, summarised in Part II, explains long-term economic growth in terms of surges of technological and institutional change driven by innovation. The ecological economics view, summarised here, explains long-term economic growth in terms of increases in primary energy inputs and improvements in the efficiency of conversion of primary energy to provide useful work and energy services of mobility, power, heating and lighting. Though there is not yet a complete theoretical picture that combines these two views, we suggest that the ideas presented here would provide useful building blocks for such a theory.

This is important to understanding the macroeconomic implications of a low carbon energy transformation. We are interested in understanding what such a transformation implies for economic growth. Unfortunately, we have argued that mainstream models of economic growth provide limited insight because of two fundamental limitations: they do not adequately represent the dynamic, dis-equilibrium nature of long-term technological and institutional change that drives economic growth, and they do not adequately represent the role of provision of energy and efficiency of energy conversion in driving economic growth.

We think that the models presented in this chapter of how changes in the availability of useful work from energy sources interact with labour and capital provision to drive economic growth provide valuable insights. However, we would argue that there are also limitations to the approach of developing models of the economy based on production functions. This approach does not capture the full complexity of the positive feedback processes and spillovers creating new economic opportunities inherent in the coevolutionary approach to understanding dynamic, historical transformation processes. In the next chapter, we discuss other theoretical ideas for understanding historical transformations that could help to bridge this gap.

Notes

1 Solow (1956).
2 See Abramovitz (1993).
3 Romer (1986); Lucas (1988). See Romer (1994) for a slightly more accessible account of the development of these ideas.
4 Brown (1994).
5 Ayres and Warr (2005); Warr and Ayres (2006); Warr and Ayres (2012).
6 Warr and Ayres (2006).

7 Denison (1979).
8 Kümmel et al. (1985); Kümmel (2011); Ayres et al. (2013).
9 Kümmel (2011).
10 Ayres and Kneese (1969).
11 Ayres and Warr (2009), p. 10.
12 Felipe and Fisher (2003).
13 Thanks to my colleague Ariel Wirkierman for pointing this out to me.
14 Warr and Ayres (2012).
15 Brockway et al. (2014, 2015).
16 Arapostathis et al. (2013).
17 Fouquet (2008, 2010, 2016a, 2016b).
18 Fouquet (2016b).
19 Fouquet and Pearson (2006).
20 Aghion and Howitt (1998).
21 Stern, D.I. (2011).
22 Stern and Kander (2012); Kander and Stern (2014).

9

INSIGHTS FOR A LOW CARBON ENERGY TRANSFORMATION

Introduction

In this chapter, we examine other historical and theoretical work that has aimed to examine the dynamic processes and role of energy in past historical industrial transformations. Combining this with our analysis of historical surges of techno-economic change, we then draw out implications for realising a low carbon industrial transformation. This draws on our work with fellow British energy economist Peter Pearson, in which we analysed what these ideas from economic historians and theorists can tell us about the potential for a low carbon industrial revolution, necessary to address the challenge of climate change.[1] This analysis addresses two questions. First, is the scale of changes needed to achieve a low carbon transformation of a similar scale to those of past industrial revolutions? Second, can a low carbon transformation deliver similar wide-scale economic benefits to those that were delivered by previous industrial transformations? We end the chapter by examining in more detail the challenge for a low carbon transformation due to declining or low net energy returns from conventional and renewable energy sources.

Lessons from economic historians

In Part II, we saw how previous surges of economic growth involved not just decreases in costs of key technologies and the services that they provided, but also significant changes in institutional systems, business strategies and user practices. But what drove those changes? As we have seen, mainstream economic theory provides limited insight into this. Our coevolutionary framework, drawing on the ideas of

Freeman and Perez, highlights key elements of the transformation processes involved in past surges of economic growth. Of course, much useful insight has also been provided by economic historians. In particular, they have analysed in detail the questions of why the industrial revolution started in Britain, and what are the feedbacks between different drivers of surges of economic growth. This has important implications for our understanding of the potential for a new surge of economic growth to be driven by low carbon energy technologies and associated institutions, in what some have claimed could be a 'new low carbon industrial revolution'.[2]

Most economic historians have identified the first industrial revolution in Britain from around the 1770s as a seismic event in world history (although it resulted from the coming together of different strands of technological and institutional change). Indeed, US-based economic historian Gregory Clark has gone so far as to say, 'The Industrial Revolution thus represents the single great event of world economic history, the change between two fundamentally different economic systems.'[3] British economic historian Tony Wrigley has argued that this represents a change from an organic economy to an industrial economy. In an organic economy, such as Britain before the second half of the eighteenth century, productivity gains are limited by the decreasing returns associated with expansion of economic activity based on energy derived from plants or animals. In other words, each additional effort put in yields a smaller amount of economic return. New knowledge, such as selective breeding, can help to improve this return, but there are fundamental limits associated with the availability of land and the ability to harness natural flows of solar and wind power. In an industrial economy, these limits are overcome by the use of coal (a stock of fossil fuel) to provide mechanical energy. This enables increasing returns to the expansion of economic activity as the ability to use coal drove further improvements in its extraction and use.[4] As we have seen, the positive feedbacks between the productivity gains resulting from new technologies, in particular the steam engine, and new institutions, particularly those supporting the free exchange of information and the development of knowledge, were crucial in driving the expansion of this new industrial economy. This led to a dramatic reduction in the cost of energy services, particularly in providing power for machinery and transport, which increased the demand for these services, stimulating further technological improvements in a positive feedback loop.

However, different economic historians vary in the emphasis they put on different drivers of these changes. British economic historian Robert Allen points to the importance of the relatively low cost of energy, due to the easy availability of water power and then coal in northern England, compared to the relatively high cost of wages.[5] Wages were relatively high, as Britain had already enjoyed benefits from being a leading trading nation, stimulating improvements in agricultural production. This meant that new innovations, such as the spinning jenny for cotton production, could be profitably adopted, as these technologies allowed higher production with fewer workers. The higher wages also meant a more highly skilled population, as workers invested in education and training, and greater demand for cotton textiles.

Thus, greater productivity in terms of output per worker led to a positive feedback loop in which greater demand for output of final products stimulated expansion of production. This fits well with the picture of positive feedbacks driving economic growth that we examined in the previous chapter. Here, Allen emphasises the role in this process of the demand for new goods and services interacting with the low cost of energy and capital (machines) compared to high labour costs in driving the adoption of energy-using machines in place of labour. Again, the overall productivity gains from this process led to an expansion in the scale of economic activity, leading to greater overall levels of employment in the textiles and other industries.

On the other hand, American-Israeli economic historian Joel Mokyr places more emphasis on the supply of knowledge.[6] He highlights the importance of the Age of Enlightenment, the revolution in liberal philosophical and political thinking in the 1700s that helped pave the way for the American and French revolutions, as well as the scientific revolution. He argues that this culture of the free exchange of ideas and ability to challenge accepted ways of thinking and vested interests helped to stimulate the production and exchange of useful knowledge. He sees that this also stimulated the development of institutions to reward innovation and entrepreneurship, rather than the appropriation of rewards by wealthy land and property owners. This aligns with the important role that Turkish-American economist Daron Acemoglu and British political economist James Robinson gave to the development of inclusive political institutions in Britain, following the Glorious Revolution of 1688, in stimulating inclusive economic institutions.[7] There has been a long history of the interactions between political and economic institutional change. The founding text of liberal market economics, the *Wealth of Nations* by Scottish economist Adam Smith, was published in 1776, the year of the American Declaration of Independence.

British economic historian Nicholas Crafts has suggested that these two explanations are complementary. He argues that Allen's emphasis on the role of economic incentives driving the surge of technological progress in the British industrial revolution should be seen in the light of the wider social and cultural change discussed by Mokyr, so that the surge 'resulted from the responsiveness of agents, which was augmented by the Enlightenment, to the wage and price configuration that underpinned the profitability of innovative effort in the eighteenth century'.[8]

General purpose technologies

Another complementary perspective on the role of technological change in stimulating economic growth is provided by the idea of 'general purpose technologies' (GPTs). In a similar way to our discussion of surges of economic growth, this idea emphasises the role of these generic technologies, such as the steam engine, electrification and ICTs, in driving periods of high economic growth. In particular, this idea highlights three key features of GPTs: they have capacity for continued innovation and falling costs; they have a wide range of applications; they stimulate complementary innovations.[9] Again, the combination of these three properties generates the positive feedbacks necessary for a stimulus to economic growth. As a GPT is developed and taken up, its costs fall, so it is used for its original purpose and stimulates new uses, which promotes further development and use of related technologies. The theory of GPTs has been criticised by some economic historians, as the influence of any particular technology is hard to identify in the historical economic data. However, those promoting the idea of GPTs, such as Canadian economists Richard Lipsey, Clifford Bekar and Kenneth Carlaw, also recognise that the institutional context, what they call the facilitating structure, is important in determining the rates of take up of any GPT and its wider economic influence. Defending against these criticisms, they argue that evolutionary economic models incorporating GPTs and their facilitating structure can form a useful bridge between mainstream economic models, such as Solow-type models, that have a very simplified representation of technologies, and the narrative descriptions of economic history that tell the story of the complex interactions between technologies, institutions and economic development.[10]

Implications for a low carbon energy transformation

So, with these further insights from economic historians and theorists of general purpose technologies, together with the insights from our analysis of historical surges of economic growth, we can return to the questions at the start of the chapter. To address the first question, these historical analyses all show that introducing new energy provision technologies, such as the steam engine and electricity, was rarely a matter of just substituting the new technology for the old in the existing structures. The systems that had developed around the old technologies had evolved to maximise the economic benefits associated with those technologies and to support the interests of those who controlled the key economic assets that delivered these benefits. So, new energy technologies only displaced old technologies when the economic benefits became clear through a process of use in particular niches and cost reductions through learning-by-doing,[11] and the political interests of those supporting the old technologies were able to be overcome. New low carbon energy technologies based on renewable energy, such as solar photovoltaics (PV) and wind power, face a similar process of demonstrating their viability in particular niches and trying to reduce their costs and improve their performance, as well as challenging the political interests supporting the current dominant high carbon fossil-fuel based energy technologies. This is necessary to show that, alongside energy efficiency improvements, low carbon energy technologies can provide similar or higher levels of useful work and energy services. A transformation to a low carbon energy system would thus seem to be of a similar scale to previous industrial transformations, and so require complementary changes in institutions, business strategies and user practices.

Before we can address the second question of whether a low carbon transformation can deliver widespread economic benefits, we need to point out three key ways in which a low carbon transition differs from previous systems changes. First, whereas previous industrial revolutions had largely been driven by individuals and firms motivated by private profit, a low carbon transition is seeking to deliver the social goal of mitigating climate change. A safe climate is an example of a common good that individuals cannot purchase for themselves. This means that purely economic incentives will not drive a low carbon revolution, unless they embody strong incentives or regulations to deliver that common good to prevent common harm. This can only be imposed by governments or, at least initially, by communities working together. Obviously, action by market actors, including firms and private investors, will be crucial to bringing about a low carbon transition. But individuals and firms, acting in their own economic self-interest, are most unlikely to deliver a clean energy system on their own.

Second, this implies a more significant role for public policy in driving the rate and direction of a low carbon transition than was the case in previous industrial transformations. Though, as we have seen, policy decisions did play important roles in previous transformations, they were generally focused on achieving economic gains for particular groups or nations rather than driving a systemic change. These policies for promoting a low carbon transition will interact with other policy objectives, including upholding national economic competitiveness at the macro level, and maintaining energy security and affordability of energy services to households and firms.

Third, the speed of achieving a low carbon transition is driven by climate change imperatives that cannot be negotiated with. The timescale for achieving a low carbon transformation by 2050 is significantly shorter than that needed for previous industrial transformations.

What would be needed for a low carbon industrial revolution?

The question of whether a low carbon transformation can deliver widespread economic benefits, similar to those delivered by previous industrial transformations, is crucial for understanding the economic and political scale of the challenge. If this were to be the case, then public policy would be just a matter of providing the right incentives to private economic actors and overcoming vested interests. If, instead, the main or only benefit is the social benefit of mitigating climate change, then there would be a bigger challenge of persuading people and firms to accept short-term costs for avoidance of catastrophic outcomes in the long term. So, what we can learn from historical studies?

Allen's emphasis on the importance of low energy and capital costs in driving technology substitution in the first industrial revolution suggests that reducing the costs of new low carbon energy technologies is key. Basic economic incentives are important, and so a high carbon price, through a carbon tax or trading scheme, and low carbon innovation policies would be needed to reduce the relative costs and promote the take-up of low carbon energy technologies. Mokyr's insights on the role of knowledge and supporting institutions suggests that training in low carbon skills and strong institutions that provide a clear direction for a low carbon transition will also be important.

However, we still need to ask whether low carbon energy technologies have similar properties to those of past general purpose technologies, or the 'key factors' that drove previous economic surges. As noted, GPTs are argued to have three features: capacity for continued innovation and falling costs; a wide range of applications; the ability to create spillovers by stimulating complementary innovations. Similarly, Freeman and Perez describe 'key factors' driving sustained economic surges as having falling costs, rapidly increasing supply and pervasive applications.[12] Low carbon energy technologies certainly have the potential for continued innovation and falling costs. The costs of these technologies, particularly solar PV panels, have been coming down rapidly in recent years with learning and experience. This has partly been driven by policies that have provided incentives for householders to fit solar panels in such countries as Germany and Spain. This created a significant market and so encouraged investment in large-scale manufacturing, particularly in China, which has driven down the costs per panel.

However, these technologies do not yet seem to have the other properties of GPTs. Though they have a wide range of applications, these are mostly providing the same energy service as would be provided by a fossil-fuel based energy technology. For example, the electricity provided by a solar PV panel provides the same service of powering devices as electricity from a large coal-fired power station or a local diesel generator. Differences can arise though in the flexibility of that provision. In areas of sunny developing countries that are not connected to electricity grids, solar PV panels can provide the cheapest and most flexible means of generating power. In industrialised countries, take-up of solar PV is being driven by government incentives, together with values of security and prestige that householders put on generating their own energy. In general, though, low carbon energy technologies are not yet creating new uses, as previous GPTs did.

Similarly, they are not yet stimulating innovation in complementary technologies or applications. We speculated that 'smart' energy systems that combine low carbon energy technologies with ICTs could have the potential to do this. For example, we are now seeing home energy management systems that enable householders to switch devices on or off from their smart phones, as well as smart grids that are able to handle two-way power flows. These types of smart systems have great potential in providing new applications, both to consumers and in improving the efficiency of production processes, but they are still in their infancy at the moment.

Overcoming carbon lock-in

The main conclusion from the history of previous technology-driven surges of economic growth, whether viewed from an evolutionary, an economic history or a GPT perspective, is that these surges were driven by a series of positive feedbacks that generally involved substituting exergy-using machines for labour and the application of knowledge to providing cheaper or higher quality services. To realise the full benefits of these innovations, systemic and complementary changes were needed in other technologies, institutional rules, business models and user practices. We have seen how new technological systems have to battle against the established technologies and ways of thinking associated with them. This is sometimes referred to as 'lock-in', and it has been argued that we are currently in a state of 'carbon lock-in', as high carbon energy technologies, e.g. coal, oil and gas, have coevolved with regulatory and economic institutions and social practices to make high carbon ways of life the norm in industrialised countries and the aspiration of many in the developing world.[13] History can help us see the scale of the challenge involved in overcoming this carbon lock-in, and provide useful insights into the processes by which past cases of systemic lock-in have been overcome, sometimes fairly rapidly. However, realising the level of change needed to overcome the current carbon lock-in and achieve the desired social outcomes will require public pressure and political will to overcome the vested interests in the current system.

Techno-economic paradigm change

Though we have focused on the role of energy technologies in driving surges of economic growth, Perez's theory of techno-economic paradigms paints a broader picture of economic change. She emphasises that a new paradigm, enabled by the falling costs and widening applications of the key factors, gradually becomes the 'new common sense' for organising economic activity. However, there is a sense in which each of the five surges also built on attributes of the previous surges.[14] So, the ICT-based surge built on the mass production and consumption paradigm of the previous oil-based surge, and, of course, the electricity systems developed in the third surge.

Perez has argued that the world should now be moving into the deployment phase of the ICT paradigm, following the dot-com bubble of the early 2000s and the 2008 financial crisis which is still reverberating. Furthermore, she believes that there should be clear synergies between the implementation of ICTs and green technologies, for example in smart energy systems, as already discussed.[15] However, she has argued that to realise the benefits of a 'green golden age' requires a more decisive split with the mass production and consumption paradigm. In her view, renewable energy technologies and associated resource-efficient innovations are 'not sufficiently far-reaching alone to revive growth'.[16] Rather, 'green' can provide a direction of stimulus for deploying ICTs as general purpose technologies across the entire economy. This would require major changes in both production patterns and lifestyles. Perez argues that this would not only turn environmental challenges 'from obstacles into solutions', it would create new sources of job creation and new opportunities for economic development in developing countries. Perez recognises that this requires overcoming the still dominant neoliberal economic 'free market dogma' and associated lack of support for strong government action, and will require pressure from social movements and opinion formers to create a 'positive sum design' between business and society.

This is an inspiring vision that would break with the logic of mass consumerism and the throwaway society by embracing 'circular economy' ideas of design of products for reuse and re-manufacturing.[17] She argues that a positive vision of social and economic change is more likely to inspire people, and so successfully galvanise that change, than framing the problem in terms of addressing the threat of climate change. We are very sympathetic to this view of a positive vision of change, but from our energy- and ecological-based perspective two questions remain. Can such a profound transformation in the core economic and political ideas of industrialised societies be accepted and implemented quickly enough to reorient global energy systems to low carbon systems, in order to prevent cata-strophic climate change? Can renewable energy systems deliver the necessary exergy inputs to provide the useful work and energy services needed, even for a highly efficient and less material intensive economy?

The potential for a fundamental transformation in societies' core economic and political ideas goes beyond what we are able to discuss in detail in this book, though we sketch some elements of this transformation in Chapter 12. We discuss the potential for a rapid energy transition in terms of changes in investment patterns and infrastructure in the next chapter. Here, though, we want to address the second question by linking the potential for a green paradigm change to the discussion of net energy returns.

New renewable energy paradigm

As might be expected, there are both positive and negative views of the potential for renewable energy to form the core energy inputs for a new techno-economic paradigm. The positive view is represented here by Australian management scholar John A. Mathews, whilst the negative view is associated with American systems ecologist and energy scholar Charles Hall and colleagues.

Mathews argues that a new form of green capitalist economy is possible, based on renewable energy technologies, circular economy ideas and reorienting the financial system to provide eco-finance.[18] He refers to this as the 'Next Great Transformation', in analogy with Polanyi's characterisation of the birth of the market economy, powered by fossil fuels, as the Great Transformation.[19] Mathews argues that the features of renewable energy technologies, i.e. that they are clean, inexhaustible, close-to-zero running costs, scalable and decentralised, means that they will become increasingly desirable as primary energy sources. He notes that China has invested heavily in developing manufacturing capacities for solar PV and wind turbines, which has helped bring the unit costs of these technologies down significantly. The key feature of renewables from this perspective is that they are based on manufacturing, rather than extraction, as is the case for fossil fuel technolo-gies. Extraction-based energy technologies are subject to decreasing returns – the cheapest and most easily accessible sources are extracted first, so that these energy sources become more difficult and expensive to extract over time. Renewable energy technologies benefit from increasing returns, since specialisation, economies of scale and learning effects in manufacturing mean that they become cheaper over time. This helps to create demand for these technologies, so stimulating more investment generating further increasing returns and cost reductions in a process of circular and cumulative causation.[20] This means that renewables will sooner or later supplant fossil fuel energy technologies. This process can be accelerated, though, by policy instruments, such as feed-in tariffs, that help to create a market for renewables, further stimulating the increasing returns and cost reductions.

Declining net energy returns

A more conservative view on the potential for renewable energy to form the core energy inputs for a new techno-economic paradigm is provided by scholars working on the concept of net energy returns. This arises from another branch of the ecological economics literature, which focuses on how much energy is needed to extract a particular primary energy source. This is usually expressed as the ratio of energy return (from a particular energy source) to the total energy invested to get that energy, or energy return on energy invested (EROI). This idea was first formulated in relation to concerns about resource depletion, or 'peak oil' – the notion that conventional supplies of oil are at or close to their peak value.[21] The concept is different to that of exergy efficiency that we examined in the previous chapter.

Essentially, EROI measures the trade-off between technological improvements, which tend to increase the net energy return, and resource depletion, which tends to decrease the net energy return, since the cheapest and most easily accessible sources are extracted first (as already noted). It is clearly important to assess which of these tendencies is likely to win out for different energy sources, given our argument that energy is a key input into economic activity. The fact that mainstream economic analysis does not assign any special role to energy as an input, and the conceptual and measurement difficulties in calculating EROI values for particular energy sources, mean that this concept has so far had little impact on mainstream economic debates. We would argue, though, that it is an important concept to consider, alongside other energy variables, such as exergy efficiency.

The argument that EROI values for conventional fossil fuels are declining and that this may have important implications for economic growth was first clearly made by Charles Hall and colleagues in 2008.[22] 'Peak oil' proponents have noted that the rate of discovery of new large oilfields has been declining since the 1960s, with no new 'mega oilfields' comparable to those in the Middle East having been discovered since then. This meant that oil companies began to move to less accessible conventional oil sources, such as the North Sea and the Gulf of Mexico, and unconventional oil sources, such as tar sands and shale oil. These are more difficult to extract and so require more energy input per unit of energy returned. Hall and colleagues estimated that the EROI for finding and production of domestic US oil dropped from greater than 100 kilojoules returned per kilojoule of energy invested in the 1930s to a ratio of 30:1 in the 1970s to less than 18:1 in the 2000s, with a similar global average at that time of around 19:1.[23] This means that more of the economic output of the economy needs to be reinvested in extracting energy, rather than being spent on basic necessities, such as food and housing, or on the discretionary spending on consumption and investment in goods and services that drives economic growth. They developed a simple model showing that if global EROI of all energy sources continues to decline to around 5:1 by 2050, then discretionary spending would decline from the current value of around 50 per cent of GDP down to only 10 per cent of GDP. Whilst the outputs of any simple model must be taken with caution, if this analysis is correct, then it would imply a significant limitation on potential future rates of economic growth.

If net energy returns are declining on conventional oil and gas sources, can unconventional fossil fuel sources or renewables provide alternatives with higher net energy returns? Unfortunately, the net energy returns from these sources seems to also be low.[24] Estimates of the EROI for bitumen from tar sands in Canada and for shale oil in the US are less than 5:1, and estimates for bioethanol produced from corn in the US are less than 2:1. Estimates of the EROI for solar PV are around 7:1, though recent analysis argues that when an appropriate comparison with grid electricity is made, the EROI for solar PV increases to 9:1 or even 19:1.[25] These variations reflect issues relating to where to draw appropriate boundaries on which energy invested and returned to include, and the fact that current manufacture of renewables requires high levels of fossil fuel inputs, making accurate comparisons difficult. As noted, though, renewables have the potential for increasing net energy returns, since they are based on manufacturing technologies to harness natural energy flows, rather than extraction of depleting fossil fuel reserves.

Energy return and increasing complexity

Finally, in this chapter, we return to insights from the broad sweep of history. American anthropologist and historian Joseph Tainter has examined the evolution and decline of complex societies, such as the Roman Empire.[26] He argues that these societies were driven by the need to expand the available energy per person, and so sought to acquire land by agreement or invasion. In doing so, they required increasing social complexity, for example in the form of new rules and institutions to manage the different interests within the society. This results in what he calls an 'energy-complexity spiral', with increasing complexity needed to solve the problems created by the use of the additional energy per person. Unfortunately, in many cases, such as that of the Roman Empire, despite the best efforts of the ruling class, the increasing complexity and cost of solving problems undermined the coherence of the society leading to eventual collapse. Drawing on Tainter's ideas, other examples of societies that had increased their complexity and overexploited their resource base, leading to societal collapse, were documented by American scientist and author Jared Diamond in his book *Collapse*.[27]

Tainter maintained that the technological and institutional breakthroughs of the industrial revolution enabled Western industrial societies to massively increase their complexity, so escaping the danger of collapse for the last two centuries. However, he contends that the declining net energy returns that societies are now facing, forcing the search for ever-more dangerous sources of fossil fuels, could be a signal of the dangers of complex over-reach of present societies.[28]

Of course, as these authors argue, these insights do not mean that collapse of industrialised societies is inevitable. Though present societies face many challenges, we also have unprecedented knowledge to bring to bear to find benign solutions. The biggest challenge is whether we can find the collective wisdom and humanity to implement these solutions. With that warning ringing in our ears, in the next chapter, we look in more detail at potential future energy paths to provide increasing access to high quality energy, whilst addressing climate change, other environmental impacts and declining net energy returns.

Notes

1 Pearson and Foxon (2012).
2 Stern, N. (2012, 2015).
3 Clark, G. (2014).
4 Wrigley (2010, 2013, 2016).
5 Allen (2009).
6 Mokyr (2009, 2016).
7 Acemoglu and Robinson (2012).
8 Crafts (2010), p. 166.
9 Lipsey et al. (2005).
10 Bekar et al. (2016)
11 Foxon (2010).
12 Freeman and Perez (1988).
13 Unruh (2000).
14 See also Schot and Kanger (2016) for a discussion of this continuity and why a second 'deep transition' may be needed.
15 Perez (2013); Mazzucato and Perez (2014).
16 Perez (2016), p. 200.
17 Ellen Macarthur Foundation (no date).
18 Mathews (2015); Young (1928).
19 Polanyi (1944/2001); Mathews (2011).
20 Mathews and Reinert (2014).
21 Murphy and Hall (2011).
22 Hall et al. (2008).
23 Hall and Klitgaard (2012).
24 Murphy and Hall (2011).
25 Carbajales-Dale et al. (2016).
26 Tainter (1988, 1996).
27 Diamond (2005).
28 Tainter and Patzek (2012).

PART IV

Future challenges

10

FUTURE ENERGY PATHWAYS AND ISSUES

Introduction

In previous chapters, we have discussed the three challenges of increasing access to high quality energy, significantly reducing human-induced carbon emissions and declining rates of net energy return. To address these challenges will require major changes in the way that energy systems are governed and in the practices for which energy is used, as well as radical technological changes. These system changes are likely to be at least as dramatic and wide-reaching as the previous energy-industrial system changes that we have examined in previous chapters.

This raises two questions, for which our historical studies can provide some guidance. First, what is needed to drive this type of systems change? Second, can this system change stimulate a similar surge in economic growth to past energy-industrial system changes? This chapter explores potential future energy paths for the UK and globally, in order to address these questions. Different visions of future pathways to a low carbon energy system and economy will have different roles for governments, private firms and wider civil society, as well as different mixes of technologies and energy efficiency improvements. The key role of energy in modern economies means that these pathways also need to ensure security of energy supplies and affordability of energy services for consumers and businesses.

We shall look at the rates of carbon emissions reduction and improvements in energy productivity needed to achieve the Paris Agreement target to keep a global temperature rise this century to well below 2°C above pre-industrial levels – that is, by how much do we need to increase the economic value of every unit of energy used and every unit of carbon emitted. We shall also consider if this target means that significant amounts of fossil fuels will need to be left in the ground, in order to avoid dangerous climate change. Finally, we shall consider how questions of reducing carbon emissions from energy are closely interwoven with the question of economic growth.

Coming to an agreement on emissions reductions

With the deployment of the fourth surge leading to a rapid expansion of oil use, alongside further increases in coal burning for electricity generation, global emissions of carbon dioxide (CO_2) grew rapidly from 6 billion tonnes in 1950 to 16 billion tonnes by 1972. Despite the oil shocks and recessions of the mid-1970s and early 1980s, these only caused temporary blips in the continuing rise of global emissions to 25 billion tonnes by 2000. The rapid economic growth in China, associated with a massive expansion of coal use for industry and electricity generation, caused a more rapid increase in emissions in the 2000s, leading global emissions to a peak of 36 billion tonnes in 2014 (see Figure 10.1).[1] Thanks to a slowing of the rate of economic growth in China and historically low rates of economic growth in the US and Europe, together with a global expansion of the use of renewable energy, global CO_2 emissions have remained at around this level in 2015 and 2016.[2]

Currently, only around half of these emissions are absorbed by additional plant growth on land and in the oceans, with the rest contributing to increasing concentrations of CO_2 in the atmosphere. Together with emissions due to cutting down forests and other land use changes, these emissions from energy and industry lead to steadily rising atmospheric concentrations. As noted in Chapter 2, these concentrations are now above 400 parts per million and rising, compared with a stable pre-industrial level of 280 parts per million. So, even if the current stable level of global CO_2 emissions is maintained, these concentrations in the atmosphere will keep on rising. This traps heat energy in the atmosphere, causing the

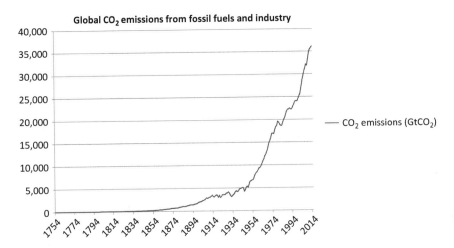

FIGURE 10.1 Global CO_2 emissions from fossil fuels and industry.
Source: Boden et al. (2017), Carbon Dioxide Information Analysis Centre.

increases in temperature across the globe, and increasing risks of extreme weather events such as hurricanes, floods, droughts and rising sea levels. Whilst it is possible and necessary to take actions to adapt to these risks, such as building higher sea defences, it is people in poorer countries and in poorer parts of richer countries who will suffer most from these risks. If increasing temperatures and extreme weather causes significant failing of crops over a prolonged period, as has been projected, then that would lead to severe social impacts, such as increased risks of conflict over resources, wars and forced migration of people.

In response to these risks, which climate scientists have been warning about since 1990, almost all countries in the world came together at the Rio Earth Summit in 1992 to sign the United Nations Framework Convention on Climate Change (UNFCCC). This committed these countries to 'stabilize greenhouse gas concentrations in the atmosphere at a level that would prevent dangerous anthropogenic [human-induced] interference with the climate system'.[3] However, the Convention did not agree any targets for reductions in emissions. These were only agreed in the Kyoto Protocol to the Convention in 1997, and then only applied to industrialised countries for the period 2008–2012. This had effect in some countries, particularly in the European Union, stimulating the development of the EU Emissions Trading Scheme that puts a price on carbon emissions from industry and electricity generation. However, other countries such as the US, Canada and Australia withdrew from, or ignored, the agreement. The main criticism made by the US government and Senate was that the Protocol did not include any targets for China or other rapidly developing countries. Though the US government was severely criticised for this by environmental groups, with hindsight, it can be argued that they had a point, as two-thirds of the growth in emissions from 1990 to 2012 came from China's rapidly developing economy.[4] Of course, much of this energy-intensive economic growth in China was driven by the production of goods and services for consumption in Western industrialised countries. This demonstrates the need for, but also the challenges of, making a global agreement to reduce carbon emissions.

A key stimulus for countries to start to make more significant emissions reductions was the economic case made in the Stern Review on the Economics of Climate Change, commissioned by then UK Prime Minister Tony Blair and Chancellor Gordon Brown, first published in 2006.[5] In this review, Nicholas (now Lord) Stern put the case for action on climate change into an economic argument that was familiar to politicians, arguing that because firms emitting greenhouse gases were not paying the costs relating to the impacts of those emissions, climate change represents the 'greatest market failure the world has ever seen'. He argued that the costs and risks of doing nothing and suffering the impacts of climate change were much higher than the costs of acting to mitigate climate change. On the basis of a range of macroeconomic models, he estimated that the costs of stabilising concentrations of greenhouse gases in the atmosphere would be around an annual additional 1–2 per cent of GDP by 2050, whereas the costs of doing nothing and suffering the impacts of climate change would be much higher, at around 5–20 per cent of GDP per year.

This economic case convinced many international politicians that climate change was a serious problem, but one that could be averted by action at a reasonable cost.

After a failed attempt to reach a global agreement at Copenhagen in 2009, the international community finally came together to reach an agreement at Paris in 2015. This agreement was subsequently ratified by the governments of over 130 countries, including the US and China, and came into force in November 2016. As noted, this commits countries to 'holding the increase in the global average temperature to well below 2 °C above pre-industrial levels and to pursue efforts to limit the temperature increase to 1.5 °C above pre-industrial levels, recognizing that this would significantly reduce the risks and impacts of climate change'.[6] This is a significant step forward, as it is the first global agreement that commits countries to a specific climate change mitigation goal, and hence was justifiably celebrated by politicians and campaigners such as former US Vice-President Al Gore and British economist Lord Nicholas Stern. However, instead of trying to negotiate an agreement on specific emissions reductions, as in previous agreements, countries were asked to propose their intended contributions to reductions. Together, these pledges for 2030 would be a reduction compared to business-as-usual, but, if fulfilled, would still currently put the world more on track towards a warming of 3°C, rather than the maximum 2°C target.[7] So, it is important that these pledges are ratcheted up in future agreements.

The key to getting a global agreement in Paris was clearly the recognition by both the US and Chinese governments that it was in their self-interest to reach an agreement. Greenhouse gas emissions in the US have fallen since 2007, largely due to the economic recession in 2008/09 and the increase in production of shale gas which has substituted for more carbon-intensive coal in electricity generation. China's rate of increase in emissions has reduced to 3 per cent in 2012, compared to over 10 per cent annual emissions increases in the 2000s. Moreover, the direct impacts of local air pollution, such as the smogs over Beijing, due to the massive increase in coal use, as well as the potential for longer-term climate change impacts on food production due to floods and droughts, helped to convince the Chinese authorities of the need for carbon emissions reductions. Under the agreement signed by US President Barack Obama and Chinese President Xi Jinping ahead of the Paris Agreement, the US committed to reducing its emissions by 26–28 per cent below 2005 levels by 2025, and China committed that its emissions would peak by 2030. With its much larger population, China's carbon emissions per person are still much lower than those of the US, though still rising at around 3 per cent per year. Now that the new US President Donald Trump has abandoned President Obama's Clean Power Plan, which would have given legal force to the US's commitments, and has announced US withdrawal from the Paris Agreement, it seems that global leadership on climate change mitigation is now passing to China.[8]

Low carbon sources for electricity generation

Given these strong commitments to reducing carbon emissions, what do they mean for the future of global energy systems? Lord Nicholas Stern, now a professor at the

London School of Economics, argues that nothing less than a new energy-industrial revolution is needed. Drawing on the work of Carlota Perez on the historical surges of economic growth, he argues that a new sixth surge from low carbon clean technology is now possible and necessary.[9] As we shall discuss in the next chapter, this is usually framed as an argument for pursuing 'green growth'.

The scale of the challenge is so great because the three fossil fuels – coal, oil and natural gas – still supply over 80 per cent of the world's energy needs.[10] These are converted into final energy forms to meet demands for energy services in three key areas: homes and commercial buildings, industry and transport. The three main conversion channels are direct burning of fossil fuels to provide heat and machine power, generation and supply of electricity, and refined transport fuels used in petrol and diesel engines.[11] Furthermore, around 2.5 billion people globally have not yet joined these commercial energy systems, instead relying on locally collected wood for heating and cooking. In order to achieve social and economic development, these people need to have access to safe and sustainable energy sources. As we discussed in Chapter 2, there are a range of other sources of energy available by harnessing natural stores or flows of energy that can be converted to provide useful work whilst giving rise to much lower carbon emissions. As we discuss further below, there are also significant opportunities to improve the energy efficiency of delivery of these energy services, so as to reduce the scale of primary energy resources needed.

Of the three conversion channels, the easiest in which to move to low carbon energy sources is electricity systems, for which there are a range of electricity generation options available. The largest current source of low carbon generation is hydro-electric power. Once in operation, hydro power is very clean, but the building of large dams often requires rerouting of rivers, potentially affecting downstream water access, and displacement of people, as was seen in the huge Three Gorges Dam in China. This means that there are practical limits to the potential for new large hydro-electric power stations. The second largest current source of low carbon generation is nuclear power. Again, nuclear power has low carbon emissions in operation, but concerns have been raised as to whether these estimates fully include carbon emissions associated with the mining of uranium and with the dismantling and disposal of waste at the end of the power station's operations. As the most accessible sources of uranium are depleted, ores with lower concentrations of uranium need to be mined, raising economic costs and carbon intensity of production. In addition, the well-known problems of accident risk, lack of safe and secure waste storage facilities and dangers of contribution to nuclear weapons proliferation are specific to nuclear power. Finally, nuclear power stations are very expensive and take a long time to build, as the UK government is currently finding out, as it tries to incentivise private companies to build the first new nuclear power stations in the UK since 1995. So, these economic and public acceptability issues will limit the potential for scale-up of new nuclear power.

The biggest contributions to low carbon electricity generation in future are likely to come from a range of renewable energy sources, including solar, wind, biomass, wave, tidal and geothermal power. Solar power can be used to generate electricity in two ways. First, in the form of solar panels, increasingly seen on the roofs of buildings, which directly convert incoming sunlight into electricity, and second in the form of concentrated solar power (CSP). In the latter, large arrays of mirrors are used to concentrate sunlight on to a central circulating fluid, such as water, to turn it into steam, which is used to generate electricity via a conventional steam turbine. CSP also has the advantage that it can be combined with heat storage in molten sands, enabling power to continue to be generated during hours of darkness. Solar panels require batteries for storage or the ability to discharge surplus electricity back to the grid. Wind power can be generated onshore or, increasingly, offshore, where larger wind turbines can be built to capture energy from stronger and more stable air flows.

Biomass covers a range of crops, including fast-growing grasses, grown to provide energy when combusted to provide electricity or heat, and both in the case of combined heat and power generation. In theory, this can be zero carbon, as the CO_2 emitted when the biomass is combusted should be compensated for by the CO_2 absorbed by the growing plants. In practice, this requires crops to be replaced when harvested, and not to substitute for existing forests or peat that absorb or are stores of more CO_2. Furthermore, the growing of biofuels should not compete with land used for food production.

Other renewable sources of electricity generation are wave power, which harnesses the energy in waves created by winds and temperature differences creating water flows in seas and oceans; tidal power, in the form of barrages or lagoons, which harnesses the energy in flows of water due to the gravitational interactions between the Earth, the Moon and the Sun; and geothermal power, which harnesses the energy beneath the Earth's surface, due to heat from the Earth's formation and natural radioactive decay.

Finally, there is the potential for continuing to use coal or gas for electricity generation if it is combined with technologies for capturing and storing the CO_2 emissions. These technologies may prove to be necessary for reducing emissions from power generation and other industrial processes if renewables cannot be ramped up fast enough.[12] However, these carbon capture and storage (CCS) technologies are also not without problems. First, though they are normally argued to remove 90 per cent of the CO_2 emissions from a gas–fired power station, recent research suggests that this may be closer to 70 per cent when looked at on a life-cycle basis and including fugitive emissions of methane (a more potent greenhouse gas) in its extraction and transportation.[13] Second, CCS technologies are still at the early stages of development and, though the individual parts of capture, transport and storage of CO_2 have been shown to work, the complete system has not yet been shown to be commercially viable without high levels of government support. Third, there is an 'energy penalty', meaning a reduction in efficiency of power stations fitted with CCS, as some energy is used in the CO_2 capture and transportation process. Fourth, the use of CCS provides a reason for the continuation of the use of coal and gas, which can then be sold as 'clean coal', for example. In the long term, it is likely that only gas- or biomass-fired generation with CCS can be used consistently with the 2°C target. The supporters of CCS, though, point to the potential for 'net negative emissions' by combining the burning of biomass, which takes in CO_2 as it grows, with CCS by preventing (most of) that CO_2 being emitted. However, this would require the significant scaling up of two not fully tested technologies.

Low carbon energy systems

Though much of the focus of efforts so far to decarbonise energy systems has been on low carbon sources of electricity generation, we need to remember that electricity only accounts for around 25 per cent of final energy demand. The majority of energy demand is for heating of homes and businesses, and for transport of people and goods. At the moment, in most industrialised countries, these demands are met by centralised networks – power grids for electricity supply, gas networks for heating, and fuel networks and filling stations for transport. Some have argued for the social and ecological benefits of much more decentralised systems, based on local renewables, including solar, wind and biomass.[14] As noted, centralised electricity generation has a maximum energy efficiency of around 40 per cent, with the remaining energy lost as waste heat. Up to 7 per cent of electricity is also lost in transmission and distribution networks. Centralised networks arose because of economies of scale and the ability to provide reliable baseload supply of electricity, mostly from burning costly fossil fuels. However, the economics of local renewables are different, in that the fuel for solar and wind generation is free once the solar panel or wind turbine has been built. This means that the marginal cost of producing the next kilowatt-hour of electricity is nearly zero. This is already challenging the profitability of centralised fossil fuel generation as local renewable generation expands in Germany, as we discuss below. Local biomass generation also provides for the potential for co-generating heat and power, with a much higher overall energy efficiency. The heat can be supplied through local district heating schemes. The Danish capital of Copenhagen already meets 97 per cent of its heating needs through a district heating network supplied by combined heat and power plants.

Renewable electricity generation sources, particularly solar and wind power, are criticised for being intermittent sources, i.e. they only provide electricity when the sun shines or the wind blows. However, this problem can be mitigated by having a range of renewable energy sources, having forms of energy storage, the ability to alter times of some types of demand and having some forms of backup generation available for those periods when there is not enough electricity being generated. This means that, at least for the foreseeable future, local renewable generation will need to be complemented by large-scale renewable generation, such as offshore wind, and other centralised low carbon generation sources.

If enough low carbon electricity can be generated from a combination of small-scale and centralised sources, it can also be expanded to a wider range of end uses.

It can then be used to substitute in the other two main conversion channels: for providing heat and machine power, and for providing motive power for transportation. For example, low carbon electricity can be used to provide heating for homes and businesses through the use of heat pumps. These work like air conditioning in reverse, by absorbing heat from the air or ground outside a building, which is cooler, and transferring it to a warmer area inside. As we saw in Chapter 2, the second law of thermodynamics tells us that heat naturally only flows from warmer to cooler places. However, the addition of electrical energy to power the condenser and compressor in the heat pump enables this natural flow to be reversed, as the entropy 'price' is more than paid for by the additional disordered heat produced in the original generation of this electricity.

Low carbon electricity can also be used to power electric vehicles, substituting for the main use of oil. Again, electric vehicles have been slow to be taken up, because of concerns over lack of range of battery-powered vehicles and lower performance than conventional gasoline- or diesel-powered vehicles. Intense efforts are now going on by new entrants, such as Tesla, as well as existing car manufacturers, to improve the performance and range of electric vehicles, so that they meet the requirements of users.

There are also other, non-electric low carbon sources for meeting heating and transportation needs. Solar power can also be used for water and space heating. Different types of biofuel are also suitable for providing heating, or, in the form of bioethanol or biodiesel, for powering vehicles in modified internal combustion engines. Again, issues with sourcing of these biofuels, as well as the generation of local air pollution, needs to be considered for these fuels. Finally, there is the potential for use of hydrogen as an alternative energy carrier. Hydrogen, produced from spare grid electricity, can be used in vehicles powered by fuel cells, which combine the hydrogen with oxygen from the air to generate electricity, with clean water (H_2O) as the only by-product. However, hydrogen's lower energy density by volume means that it needs to be compressed for use in vehicles, reducing its energy benefits.

So, this means that there are a range of different possible low carbon pathways, depending on the sources available in a country and whether the form of governance of the energy system is more market-led, more government-led or more civil society-led.[15] This will require a combination of incentives, including a price on carbon emissions and support for innovation and deployment of low carbon energy sources.[16] Some sources, such as nuclear power, coal or gas with CCS, concentrated solar power and offshore wind, are more suited to current centralised electricity systems. Other sources, particularly solar panels and biomass combined heat and power systems, can be used at a much smaller scale to provide electricity and heat much closer to end users in decentralised systems. This has potential advantages, though it would require further institutional changes from current centralised systems.[17] As we have seen, in previous surges, the advantages of economies of scale associated with centralised energy systems proved a decisive factor in cost reductions for commercialisation of these technologies. However, we are now seeing rapid cost reductions and learning improvements in the manufacture of solar panels and wind turbines. In developing countries, without an existing stable centralised electricity system, this can already make decentralised solar and wind power commercially viable. In industrialised countries, the fit between these localised power sources and mobile information and communication technologies could also make a more decentralised route appealing.

Realising a low carbon transition

Though not all of these low carbon technologies are yet ready to be commercially viable without further performance improvement and cost reductions, the range of these options illustrates that achieving a low carbon is not primarily a technological challenge. Given the risks of unabated climate change, in an economically rational world, we would invest in researching and deploying these low carbon energy options to bring down their costs, and so maintain secure and affordable energy supplies whilst mitigating climate change. However, as we have seen in our studies of historical techno-economic surges, technologies coevolve with institutional rules, business strategies and user practices, as well as with ecological systems, resulting from choices made by individuals and organisations with bounded rationality.

The importance of this coevolutionary perspective for a low carbon transition was first spelled out by American business and sustainability scholar Gregory Unruh in a ground-breaking paper in 2000.[18] Building on the complex systems ideas of Brian Arthur,[19] as well as the evolutionary economics of Nelson and Winter[20] and others (see Chapter 3), Unruh argued that current fossil-fuel based energy systems are in a state of 'carbon lock-in'. Investments in the development and deployment of key technologies for energy conversion and use in these systems, particularly the coal-based electricity system and the oil-based transportation system, meant that these technologies benefited from positive feedbacks to their adoption. This means that the more these technologies are adopted, the more likely they are to be further adopted. For example, economies of scale and learning effects mean the cost of technologies, such as coal-fired power stations and gasoline-powered vehicles, come down as they are adopted. Moreover, as both producers and users become familiar with these technologies and confident in their performance, they become more likely to adopt that technology in the future. As these technological systems grow, network effects reinforce their growth, as the more other users have a compatible technology, the more it is in the next user's interests to adopt that technology. For example, similar types of vehicles generate repair shops and second-hand dealerships that reinforce the benefits of choosing a compatible option.

As Unruh pointed out, it is not just the fossil fuel technologies that have benefited from these types of positive feedbacks or increasing returns to adoption. The formal and informal institutional rules that govern these systems, for example regulatory frameworks or social norms, also exhibit similar types of positive reinforcement to their adoption. As expounded by American institutional economist Douglass North and political scientist Paul Pierson, economic and political institutions demonstrate increasing returns to adoption.[21] For example, a new regulatory framework takes political effort and negotiations to put in place but, once in place, firms and users learn to act within these rules, benefit from this learning and coordination with others, and create expectations and plans based on these rules remaining in place in the future. So, the more established a framework is, the more likely it is to remain in place. Moreover, as Unruh argued, the technologies and institutions coevolve, so that the coal-based electricity system and the oil-based transportation system are further reinforced by the positive feedbacks to the institutional rules supporting these systems. For example, actors, such as large firms, that benefit from the current institutional rules, such as the regulatory frameworks governing electricity networks, will often act to try to prevent those rules being changed. This was seen in Germany in the 1990s as dominant large energy firms tried to prevent the passing of the 'feed-in' law that would provide a guaranteed price to new renewable generation.[22]

Though these types of positive feedbacks can create lock-in to existing systems, they also mean that, once started, systems change can occur more quickly than would otherwise be the case. A key factor here is the role of positive expectations. If enough firms and users start to believe that a low carbon future is the only feasible and desirable pathway, then they will start to modify their choices and behaviours to adapt to this. So, a crucial role for governments is to create positive incentives and a credible long-term framework for a low carbon transition, so as to convince firms and users to modify their expectations in this way.

In Germany, with support from a coalition of renewable energy firms and public environmental groups, the blocking efforts of the large energy firms were overcome and the feed-in tariff to support small-scale renewable generation was introduced and extended.[23] Along with similar economic incentives in other European countries, such as Spain and Denmark, this helped to create a market for rooftop solar panels and other renewable energy technologies. In turn, the availability of this market helped to convince manufacturers in China to invest heavily in the manufacture of these solar panels. This created economies of scale and learning effects leading to dramatic reductions in the cost of solar photovoltaic technology, which, of course, increase their attractiveness to users. Thus, in this case, changes to institutional rules on the demand-side helped to stimulate the increase in supply of a low carbon energy technology.

Carbon bubble

Another aspect of lock-in to the current fossil-fuel based energy system has recently been identified and begun to be challenged by social and environmental groups. It is clear that high levels of investment will be needed to realise the development and implementation of low carbon energy systems. Estimates from the International Energy Agency (IEA) and others propose that, globally, investments in low carbon energy technologies and energy efficiency measures will need to rise to nearly \$2 trillion per year in order to transform energy systems to meet the 2°C Paris target.[24] Though this is obviously a large sum, arguments that this level of finance is not available can be contested. An eye-watering sum of over \$70 trillion is invested annually by the global financial system in productive and financial assets.[25] So, the challenge is to divert a relatively small proportion of these investments towards low carbon and energy efficiency investments, by making these economically as well as environmentally desirable.

The large investors within the financial system need to achieve acceptable rates of return on their investments. For example, the pension funds that manage our pension pots need to generate returns, in order to be able to have the financial resources to pay our pensions. Until recently, large energy companies, particularly those involved in the extraction and supply of oil, have been seen as safe and reliable investments by these large investors, and so are often featured heavily in their investment portfolios. Many new renewable energy companies do not have similar track records and so are seen as more risky investments.[26]

Moreover, expectations of the future play out in another important way. Part of the value of large fossil fuel companies lies in the reserves of oil and gas that are on their books, ready to be extracted for sale in future years. However, recent research has shown that the total amount of fossil fuels that can be burned whilst remaining within the 2°C target is much smaller than the total amount of known reserves.[27] In turn, this means that much of the reserves that have been booked in this way by large oil and gas companies can never be burned, if we are to stick to these climate targets. This has been referred to as a 'carbon bubble', because these unburnable reserves inflate the stock market value of these companies.[28] If enough investors start to have expectations that governments and citizens will act to enforce climate targets, then this bubble will burst, leading the value of these companies to crash. This need not trigger another global financial crisis, if the investment transition is done in a managed way, as this would be a case of transferring investment to new productive investments in low carbon technologies, rather than an overall reduction in investment. However, the fossil fuel companies stand to lose billions of dollars unless they accept that they need to become part of the transition.

Scale of emissions reductions needed

There are many scenarios produced by international agencies and academics that show how a transition to a low carbon energy system could be achieved, through a combination of low carbon electricity from renewables, nuclear power and coal and gas with carbon capture and storage, decarbonisation of heat and transport through electrification or biofuels, and enhanced energy efficiency measures. However, these scenarios often struggle to communicate the scale of the challenge involved. A recent paper in the leading journal *Science*, led by Swedish environmental scientist Johan Rockström, attempted to do this.[29]

Rockström and colleagues examined what would be needed to meet the Paris target of keeping global temperature increase to within 2°C (with a 66 per cent probability), and note that it is indeed a 'herculean task'. They argue that emissions reductions need to follow a 'carbon law' of halving CO_2 emissions every decade, from the current value of 40 gigatons per year down to 24 by 2030, 14 by 2040 and 5 gigatons per year by 2050. This would not be a law imposed by a supra-national authority, but they argue that it could provide guidance for the levels of innovation and deployment of low carbon technologies and energy efficiency measures needed. As we have seen, expectations about future developments can play a key role in guiding current decision making. Rockström has explained that the analogy that he had in mind is with 'Moore's law' that helped to guide the development of the semiconductor industry.[30] As we discussed in Chapter 7, in 1965, the founder of Intel, Gordon Moore, predicted that the number of components on an integrated circuit, and hence its computing power, would double roughly every two years, and innovations by engineers and computer scientists helped make this 'law' hold true over the next 40 years. This type of continuously reinforcing emissions reductions is necessary because it is the cumulative emissions that determine atmospheric concentrations of greenhouse gases that lead to climate change.

However, this scale of emissions reductions would not come easily and would require support for research and development and rapid deployment of low carbon technologies and energy efficiency measures, a carbon price starting at $50 per ton of CO_2 and phasing out of subsidies for fossil fuels. Even then, this would require technologies to reduce the amount of CO_2 in the atmosphere, either by combining burning biomass with CCS to store the carbon underground, or by directly sucking carbon out of the air. Unfortunately, both these so-called 'negative emissions' technologies are risky and unproven at scale.

This scenario does not discuss the implications for economic growth of this change, but it seems implicitly to follow a 'green growth' trajectory to stimulate the levels of investment needed to realise this transformation. In contrast, Kevin Anderson and Alice Bows, from the Tyndall Centre for Climate Change Research at the University of Manchester, have argued that these types of scenarios are too entwined with orthodox economic thinking and require unfeasible levels of technological progress. They conclude that 'climate change commitments are incompatible with short- to medium-term economic growth',[31] reflecting 'degrowth' arguments (see Chapter 11).

Speed of energy transitions

Rockström and colleagues argue for a rapid energy system transition by analogy with the rapid transition in information technologies associated with rapid improvements in integrated microchips. In that case, these advances were built on basic physical knowledge and it was clear that technological improvements related directly to improvements in the service of information processing. In energy technologies, as we saw in Chapter 8, technological improvements have dramatically reduced the cost of energy services of heating, power and lighting, which in turn has stimulated increasing consumption of these services. However, this still leaves open the question of whether a transition to low carbon energy systems can be achieved at the rapid pace needed to meet climate change mitigation targets, and what will be the impacts on costs and consumption of energy services.

Energy historians and systems analysts, such as Czech-Canadian academic Vaclav Smil and Austrian research scholar Arnulf Grubler, have argued that energy system transitions typically proceed slowly.[32] Grubler argued that the transition from traditional biofuels to coal-based economies took over 100 years in most European economies, and the subsequent transition to predominantly oil-based economies took around 60 years.[33] This relates to the challenge of overcoming the lock-in of systems in relation to technologies, institutions, business strategies and user practices, as we have already discussed. Typically, new energy technologies are initially more expensive than the existing alternatives and only deliver advantages in a few niche applications, to users who may be willing to pay a premium for the additional service delivered. These innovative first movers, together with investors willing to take risks to generate high rewards to scale-up the technology, help bring down the costs of the technology, and promote more widespread deployment.

However, UK-based academic Benjamin Sovacool pointed out that there are examples in some countries of certain technologies scaling up much faster than this.[34] These included the transition from oil-based electricity generation and domestic heating systems to coal- and natural gas-based combined heat and power systems in Denmark in the 1970s and the scale-up of nuclear power in France in the 1970s. Both of these were stimulated by government action following the oil price crisis in 1973, and took less than 11 years. The Canadian province of Ontario took a similar length of time to phase out its coal-fired electricity generation from 25 per cent in 2004 to 0 per cent in 2014, driven by climate change and local air pollution concerns. In response, Grubler and colleagues Charlie Wilson and Gregory Nemet argued that whilst this type of rapid technological substitution may occur in particular country cases, transitions in whole technological systems, which consist of complex sets of interrelated technologies, infrastructures and institutions, usually take many decades.[35] For example, the transition to modern transport systems took 80 years in both the capitalist US and the USSR centrally planned economy. Roger Fouquet argued that a fall in the cost of energy services, due to technology and conversion efficiency improvements, or an increase in the quality of services provided, such as cleaner burning in the home, were crucial to past energy system transitions.[36] He also pointed out that incumbent industries often react to the challenge of new technologies by improving their performance. This is referred to as the 'sailing ship' effect, after the improvement in performance of sailing ships in response to competition from steamships.

These concerns are relevant particularly for a transition to renewable energy systems. Though the costs of renewable energy technologies have come down significantly, thanks to policies to support their use in countries such as Germany and manufacture in countries such as China, they still face formidable challenges in moving to widespread adoption. First, this requires transitions in whole technology systems for heating and transport services, as well as electricity provision. At the moment, renewable electricity generally provides the same service as fossil fuel electricity, and renewable heat and transport options provide lower quality services to most users. One exception may be the case of decentralised electricity generation technologies, especially in countries where grid electricity is not available or not reliable. This suggests there may be an advantage for developing countries to 'leapfrog' straight to decentralised renewable energy systems.[37] However, in industrialised countries, the advantages of present centralised systems and the power of incumbent firms within these systems to block change make this much more difficult. Sailing ship effects can also be seen in fossil fuel technologies, such as the production of shale oil and gas by fracking, driven by technological improvements and high prices of conventional oil supplies, as well as by incumbent energy firms responding to potential competition from renewables.

Both optimists and pessimists agree that whilst these historical cases can provide useful guidance, they cannot be used to predict future outcomes.[38] In the end, the future depends on the choices of the range of large and small actors within energy systems and the political pressure put on governments to act. The Paris Agreement to reduce greenhouse gas emissions could mark a turning point when nations start to act, individually and collectively, to institute a transition to a low carbon global energy system. Whether this, and subsequent policy actions to promote this, will succeed depends partly on expectations. If enough individuals and firms believe that a low carbon energy transition is going to happen, then they will act to make it happen.

Improving energy productivity

Whilst a transition from high carbon to low carbon primary energy sources is essential for meeting climate change targets, a significant improvement in the efficiency with which these primary energy sources are converted to provide useful energy services for households and businesses will also be necessary. The IEA estimates that almost half of the reductions in CO_2 emissions from a business-as-usual trajectory (which would put the world on track to 6°C of warming) to a 2°C scenario (2DS) by 2050 would have to come from energy efficiency improvements.[39] As these scenarios assume a continuing high rate of economic growth, averaging around 3 per cent (with less than 2 per cent annual economic growth in richer countries, and 4–5 per cent annual economic growth in developing and emerging economies), this implies a significant improvement in energy productivity, i.e. the economic value provided by each unit of energy used. In the IEA's 2DS scenario, global final energy demand only grows from 390 exajoules (EJ) in 2014 to 455 EJ in 2050. This implies that energy productivity (GDP/energy) improves at an annual rate of just under 3 per cent.

How could such a radical improvement in energy productivity be achieved? A useful way to think about this is to separate these changes into two components: energy efficiency of services, and GDP productivity of services, as follows:[40]

$$GDP/energy = service\ level/energy \times GDP/service\ level$$

The key here is that people and business want the services that energy provides, including mobility of people or freight, heating and lighting in buildings, and outputs, e.g. tonnes of steel, in industry. Improvements in energy efficiency of services would allow these to be provided with much lower physical amounts of energy, by changes in technologies or ways of organising production and consumption. Improvements in GDP productivity of services would enable more economic value to be created for a given level of energy services, through structural or behavioural changes in economic activity.

In providing mobility, this would imply improving the fuel economy of all vehicles, switching to more efficient electric vehicles (alongside decarbonisation of grid electricity) and encouraging people to use public transport instead of private vehicles, as well as structural changes, such as moving to more compact cities and using video conferencing to reduce travel demand. In industry, this would imply improving the energy efficiency of industrial processes, making use of waste heat from these processes, and structural changes, such as moving to a circular economy, in which products are designed for recycling, reuse or remanufacturing to reduce the amount of new products needed.[41] In buildings, improving the design of new buildings and retrofitting old buildings could reduce the amount of energy needed for heating, lighting and ventilation by three- or four-fold for the same floor area, as well as the use of smart meters and more efficient appliances to help people use less energy, whilst maintaining levels of comfort and service delivery.

So, there is a significant range of opportunities for improving the energy productivity of economies by implementing energy efficiency improvements and pursuing structural changes to reduce the energy intensity of economic activities. This would create opportunities for profitable innovation, combining new technological opportunities with new business models for creating and capturing value.[42] These would need to be combined with incentives for the innovation and deployment of the range of low carbon technologies that we discussed previously. Different countries can choose to make use of different mixes of low carbon and energy productivity improvements, depending on their current resource and skills base and state of economic development.

Reorienting economies to low carbon and energy productive pathways

However, at least three challenges remain to be overcome in reorienting national and global economies towards low carbon and energy productive pathways. First, how to prevent improvements in the energy efficiency of economies being overwhelmed by increases in levels of consumption? Second, how to reorient the global financial system so as to provide the levels of investment needed to realise these opportunities? Third, how to galvanise the political will to overcome the lock-in of current high carbon technological and institutional systems, and associated economic ways of thinking that support these? We now discuss each of these issues in turn.

Mitigating rebound effects

As we have seen from our historical analyses, past economic surges have been driven by technological and institutional innovations that have reduced the cost of key inputs, created opportunities for spillovers to new uses, and enabled or enhanced meeting human needs and wants. As we noted in Chapter 5, a key aspect of this in relation to potential ecological and resource limits is the so-called rebound effect. Improving the efficiency of the provision of a service, say mobility, implies that less energy and other inputs are needed to provide that service, but it also effectively reduces the cost of the service. In turn, this tends to increase the demand for that service. This means that an improvement in energy efficiency of delivery of a given service will not lead to the expected reduction in energy needed, as some of the saving will be 'taken back' as increased consumption. This is known as a 'rebound effect'. In fact, there are different types and scales of rebound effects.

The most straightforward direct effect is in the increased consumption of the same service. For example, making cars more fuel efficient implies that it is cheaper for people to travel. This means that they may decide to travel further or more often. For example, if the average fuel economy of cars in a fleet were to be improved from needing 100 litres per 100 kilometres to only 80 litres per 100 kilometres, and an average car owner travels 200 kilometres per week, then this would seem to imply that $200 - 160 = 40$ litres less fuel would be needed. However, if each car owner decides to travel 210 kilometres per week as a result of the reduced cost of motoring, then 168 litres of fuel will be needed, leading to a reduction of only 32 litres. The rebound effect is usually measured as the percentage difference between the expected and actual energy savings, in this case 20 per cent. So, energy is still saved as a result of the efficiency improvement, but less energy is saved that would naively be expected. There may also be indirect rebound effects. For example, instead of using the money saved to drive further, the car owner may decide to use this saving to take an additional holiday abroad. This was demonstrated with unintentional irony by UK supermarket giant Tesco in a promotion offering its customers air miles in exchange for buying energy efficient light bulbs, marketed as 'turning lights into flights'.[43] The size of this type of rebound effect depends on the relative energy intensity of different activities. In an article partly stimulated by the Tesco campaign, Mona Chitnis from the University of Surrey and colleagues estimated the size of the direct and indirect rebound effects from typical measures to improve energy efficiency of UK housing, such as installing wall and loft insulation and switching to LED lightbulbs.[44] They found that these measures led to rebound effects of 5–15 per cent, particularly due to households choosing to spend some of the savings on more energy and carbon-intensive uses, such as heating the building. Other studies have found similar scales of direct and indirect rebound effects, indicating that energy efficiency measures do lead to energy and carbon savings, albeit smaller savings than might at first be expected.

More controversial are efforts to estimate the size of so-called macroeconomic rebound effects. These arise from economy-wide actions to improve energy efficiency that may stimulate new large-scale economic activity. As noted in Chapter 5, this possibility was first pointed out by Jevons in 1865, who argued that more efficient steam engines stimulated increased demand for coal and so increased, rather than decreased, coal consumption.[45] This is sometimes known as the 'Jevons paradox' or 'backfire'. However, we can now see that this is a normal part of economic development – increasing efficiency of conversion of primary energy to useful work is a key driver of economic growth. This creates a tension between environmental and economic goals. From an environmental perspective, we would hope that widespread take-up of energy efficiency measures by households and industry would reduce energy consumption and so contribute to overall reductions in carbon emissions. From an economic perspective, though, widespread take-up of energy efficiency measures could stimulate economic growth by enabling new economic activity to occur, beyond just substituting from one end use to another. This is a question of scale and aggregation. Put simply, replacing the incandescent light bulbs in your home with LED lights is not going to stimulate economic growth, but finding a more efficient way of producing lighting might.

In order to realise the potential for efficiency improvements to be environmentally as well as economically beneficial at a large scale, additional policy measures will be needed. In the context of reducing carbon emissions, an overall cap on the carbon emissions within an economy would limit the ability for efficiency savings to be taken back in more carbon-intensive activities, such as flying. Thus, energy productivity improvements should be seen as complementary to moves towards low carbon energy provision. More generally, efforts to improve energy productivity should be seen in a wider way. In the present political context, if energy efficiency improvements also help to stimulate economic growth, then that would be seen as beneficial, provided that it is done consistently with reducing carbon emissions and other environmental goals, such as by having a carbon cap. In the next chapter, we will take up the question of the extent to which it will continue to be possible over time to stimulate both economic growth and carbon reductions by these types of energy productivity improvements and low carbon technology innovations.

Reorienting financial systems

As we have already discussed, global financial systems have not yet woken up to the challenge of providing investment for a low carbon transition. The value of fossil fuel assets does not reflect the fact that much of these reserves will need to be left in the ground, and investment in low carbon options is still often seen as too risky. In work with colleagues Stephen Hall at the University of Leeds and Ronan Bolton at the University of Edinburgh, we identified four structural barriers to the reorientation of the investment community towards renewable energy options.[46] First, renewable energy is still seen as an unfamiliar and immature investment sector, with a lack of agreed benchmarks and standards. Second, renewable energy investments typically yield returns in the long term, rather than the short term, and investors may not be willing to tie up their money for 10–15 years to get a return. Third, even though institutional investors like pension funds might want safe long-term investments, the incentives for fund managers are often to deliver returns in the shorter term to earn bonuses and high salaries. Fourth, there is a lack of secondary market investment vehicles for selling on these long-term assets. We argued that these four factors need to be addressed, in order for the investment community to be fully engaged in a low carbon transition. We argued that lifting the lid of the 'black box' of energy finance in this way means that it is better understood as a context-dependent adaptive market, rather than as the efficient market of mainstream economic theory.

Despite these structural barriers, the Inquiry into the Design of a Sustainable Financial System by the United Nations Environment Program (UNEP) identified a 'quiet revolution' going on around the world in the creation of new financial and investment practices more aligned to the environmental and social goals of sustainable development.[47] For example, publicly funded international development banks are now often willing to take on the first tranche of risk in a renewable energy project, so as to encourage more risk-averse private investors to also invest in the project. Whether these green shoots will lead to a more fundamental transformation of financial ecosystems needed to deliver a sustainable green transformation remains to be seen.[48]

Generating political will

As we have seen, a transition to a low carbon energy pathway will require a transformation in institutions, practices and business strategies, as well as in technologies, to overcome the lock-in of the fossil-fuel based energy system that still currently provides the vast majority of our energy service needs. This transformation is then of a similar scale to those of the historical transformations in technological and economic systems that drove past surges of economic growth. Those transformations clearly required political action at various stages, for example to open up free trade or to develop the regulatory frameworks around new technologies. There were also vested interests in current systems to be overcome. However, in those cases, the benefits to individual innovators and entrepreneurs largely aligned with the perceived national economic benefit. As soon as new technologies demonstrated that they could generate large economic returns, policy makers were inclined to support their further development and deployment. The crucial difference in the case of low carbon technologies is that their benefits are currently largely perceived as being the social benefit of mitigating climate change and the long-term return of avoiding catastrophic climate impacts, rather than national economic benefits. This tends to make policy makers hesitant to act, as they think that voters are primarily driven by the perception of national economic benefits that create opportunities for them in terms of jobs and livelihoods.

So, an assessment of the potential economic benefits of a low carbon energy transition, including the prospects of net job creation, would be likely to enhance political support. There would seem to be a strong case, based on the historical parallels, that the investment in innovation and deployment of low carbon energy technologies and energy productivity improvements could deliver an economic boost, including significant creation of new jobs.

Lessons from the energy transition in Germany

The other aspect to consider is that systemic change could take a long time to achieve. Arguably, the country that is most advanced is Germany, as it has been formally pursuing an 'Energiewende' or energy transition since 2010.[49] This includes a target of reducing Germany's greenhouse gas emissions by 80–95 per cent by 2050 and achieving 60 per cent of its final energy consumption from renewables by 2050, and it has widespread public support. This support dates back to public opposition to nuclear power from the 1970s, which intensified with the Chernobyl nuclear power accident in the Soviet Union in 1986. This created an alliance of different groups, including environmental groups and the coal industry, which lobbied for a phase-out of nuclear power in Germany. This was accepted by the Social Democrat–Green government in 2000, which put in place strong support for an expansion of renewable electricity generation, including a feed-in law to provide a guaranteed price and require distribution companies to connect renewables to the grid. The nuclear phase-out was due to be delayed by the Conservative-led government of Chancellor Angela Merkel, but, following the Fukushima nuclear accident in Japan in 2011, it was reinstated with a target of an end to all nuclear generation in Germany by 2022. Continued strong support for the expansion of renewables has meant that, by 2016, 30 per cent of Germany's electricity was generated from renewables on average over the year, with nearly 100 per cent of its power demand met from renewables on Sunday 8 May 2016.

However, this illustrates the scale of the challenge to be achieved in three ways. First, this intermittency of generation from solar and wind power means that some form of back-up or storage capacity, or management of demand, is necessary to maintain security of supply. At the moment, this is mainly provided by conventional coal- or gas-fired generation. However, it may be uneconomic for these conventional generators to only supply for a relatively low proportion of the time. Already, big energy firms owning gas-fired generation are complaining that renewable operators are 'stealing their profits' by supplying at peak times, when conventional generators would normally make a significant proportion of their revenue.[50] The second difficulty is that, in the short term, renewables are currently only scaling up fast enough to replace the low carbon nuclear generation that is being retired, so Germany's overall rate of carbon emissions reduction has slowed. The more difficult test is to phase out coal-fired generation, which is the most carbon intensive and still accounts for 43 per cent of Germany's electricity generation. In general, most governments are supportive of innovation and deployment of new renewable energy technologies, but are much more reluctant to advocate the phase-out of old technologies, such as coal-fired generation, that are linked to jobs and communities in particular areas.[51] The third challenge is that Germany's Energiewende is still largely an electricity transition, with relatively little progress on the decarbonisation of heat and transport. As we have said, there is potential for electrification of heat and transport services, for example through the use of heat pumps and electric vehicles, but as these require active choices by users to adopt, this will be more difficult to achieve.

Going forward

There are two potential responses to the challenge of generating political support for a low carbon energy transition. The first is to argue that the levels of investment in low carbon energy technologies needed to achieve mitigation targets can stimulate a new surge of economic growth. This is broadly the argument of those who support 'green growth'. The second is to argue that we need a more fundamental revision of the priorities of our economic system in industrialised countries, so that we put less emphasis on levels of production and consumption and more emphasis on improving wellbeing and social equality. This is the argument of those who support 'degrowth' or 'post-growth'. We shall examine these arguments in more detail in the next chapter.

Notes

1 Boden et al. (2017). Note that these values for CO_2 emissions include those from cement production and gas flaring.
2 Canadell et al. (2016).
3 United Nations (1992).
4 Clark (2012).
5 Stern, N. (2007).
6 United Nations (2015a).
7 United Nations Environment Programme (2016).
8 Stern, N. (2017).
9 Stern, N. (2012, 2015).
10 IEA (2016b).
11 Grubb et al. (2014).
12 Jaccard (2005).
13 Hammond and O'Grady (2014).
14 Alstone et al. (2015).
15 Foxon (2013).

16 Stern, N. (2007).
17 Realising Transition Pathways Engine Room (2015).
18 Unruh (2000).
19 Arthur (1989).
20 Nelson and Winter (1982).
21 North (1990); Pierson (2000).
22 Jacobsson and Lauber (2006).
23 Stenzel and Frenzel (2008).
24 IEA (2014).
25 Boston Consulting Group (2016).
26 Hall et al. (2017).
27 McGlade and Ekins (2015).
28 Carbon Tracker Initiative (2011).
29 Rockström et al. (2017).
30 Plumer (2017).
31 Anderson and Bows (2012), p. 640.
32 Smil (2010).
33 Grubler (2012).
34 Sovacool (2016)
35 Grubler et al. (2016).
36 Fouquet (2016a).
37 Goldemberg (1998).
38 Sovacool and Geels (2016).
39 IEA (2016a).
40 Vivid Economics (2017).
41 Ellen Macarthur Foundation (no date).
42 Foxon et al. (2015).
43 See Gillespie (2009).
44 Chitnis et al. (2013).
45 Jevons (1865).
46 Hall et al. (2017).
47 United Nations Environment Programme (2015).
48 Naidoo (2016).
49 See Schreurs (2015); Morris and Pehnt (2012/2016).
50 Citi GPS (2013).
51 Turnheim and Geels (2012).

11

ECONOMIC GROWTH AND BEYOND

Introduction

In this chapter, we step back to look at the challenges that we have identified in the context of recent debates about the desirability of continuing the present focus on achieving high rates of economic growth. Broadly, there are two views put forward by those who agree that we need to urgently move to an economy based on low carbon energy sources through the type of transition pathways that we looked at in the previous chapter. On the one hand, advocates of 'green growth' argue that maintaining high levels of economic growth is not only necessary for the investment needed for a low carbon transition, but that this investment is likely to be a significant driver of economic growth – the main 'growth story' as Nicholas Stern puts it.[1] On the other hand, more radical critics argue that the current paradigm of rising economic growth as the main source of human wellbeing is inconsistent with addressing climate change and maintaining the natural systems on which all human activity depends, and so argue for some form of 'degrowth'.[2] In practice, both of these approaches would imply a significant reorientation of current economic activity in both industrialised and emerging economies, but the differences in the underlying world views are deep and philosophical. A few voices have begun to argue that this polarised debate is not helpful, and that, though economic growth has delivered real and significant improvements in human wellbeing over the last three centuries, it is time for political economy to move beyond a focus on economic growth and pursue an 'agrowth' strategy instead.[3]

Here, we try to map out these debates and the strengths and weaknesses of the different positions. In general, these debates have not taken into account the historical and theoretical insights into the relations between energy use and economic growth that we have mapped in this book, so, hopefully, these insights could inform these debates. However, the differences in world views mean that there is likely to be no simple resolution of the argument, one way or another.

Limits to growth

Concerns about the tensions between rising economic growth needed to provide for developing and growing populations date back at least to British economist Thomas Malthus, who first published his famous book *An Essay on the Principle of Population* in 1798.[4] He argued that, unchecked, populations would tend to grow geometrically (1, 2, 4, 8, 16, etc.), whereas food supply would only grow arithmetically (1, 2, 3, 4, 5, etc.) so that population growth would rapidly outstrip available food supply, leading to population crashes through disease or war. For example, if a couple had four children surviving to reproducing age, the population of that country would double in around 25 years, whereas limits on land area and crop production would prevent food supply from growing sufficiently in response. The debate between 'pessimists', who point to limits to the availability of food and other resources on a finite planet, and 'optimists', who argue that such limits stimulate human ingenuity to overcome these limits through innovation, has raged ever since. In general, the optimists have seemed to have had the better of these arguments, pointing to the increases in energy and food supply through the types of technological and institutional innovations that we have discussed in this book. For example, the so-called 'green revolution' in the 1960s of the development of new high-yielding wheat and rice crops, together with increasing use of fossil-fuel based fertilisers, helped to avert starvation for millions of people in India and other developing countries. These innovations have enabled the world population to grow from less than 10 million people 10,000 years ago to just under 1 billion people in Malthus's time to over 7.4 billion people in 2016. Nevertheless, the pessimists argue that we have already overshot the physical limits of the productive capacity of the planet, so that the ghost of Malthus will come back to haunt us.

The debate was rejuvenated by two developments in ideas in the early 1970s. First, US scientists Paul Ehrlich and John Holdren published an equation that tried to disaggregate human impacts on the environment, known as the *IPAT* equation.[5] This states that

$$I = P \times A \times T$$

i.e. that, at a given time, any impact I on the environment, such as carbon dioxide emissions, is the product of population P, affluence A, usually measured by GDP per person, and technology T, measured as emissions per unit of GDP. In general, population and affluence have grown historically, so impacts are mitigated by improvements in technology, which reduce emissions (or other impact) per unit of GDP. This is sometimes referred to as 'decoupling' of economic growth from environmental impact. If technological improvements only partially offset increases in population and affluence, so that environmental impact continues to rise, but at a slower rate, then this is 'relative decoupling'. If technological improvements completely offset increases in P and A, causing a reduction in environmental impact, then this is 'absolute decoupling'. As we discuss below, historically, we have seen a relative decoupling of carbon emissions from economic growth, i.e. growth has become less carbon-intensive, but we need large absolute decoupling of carbon emissions if we are to achieve climate change mitigation targets.

The second development was the publication in 1972 of the book *The Limits to Growth* by US and Norwegian environmental and systems scientists Donella Meadows, Dennis Meadows, Jørgen Randers and William Behrens, then working at MIT.[6] They used early computer models of the relations between global economic systems and natural systems, analysing the feedbacks between resource use, industrial activity and the ability of ecosystems to absorb pollution. They showed that these models could reproduce past trends from 1900 to 1970, and used them to simulate the future impacts of human activity under different assumptions or scenarios. Their main scenarios projected that human activity would overshoot ecological limits, causing collapse of the global economic system by around 2040. Meadows and colleagues were criticised for assuming that technology only progressed in a linear fashion whilst population and industrial demand for resources grew exponentially and, perhaps unfairly, they were characterised as 'Malthus with a computer'.[7] However, their standard run projections for resource use and emissions have turned out to correspond well to actual use and emissions since 1970,[8] implying that the world could still be on a pathway to overshoot. Again, the question is whether the rate of technological improvements that have so far prevented collapse, at least in Western industrial countries, can be maintained indefinitely.

The oil price shocks in 1973 and 1979 that we discussed in Chapter 7 lent credence to the warnings of overshoot provided by the MIT team, and this and similar analyses helped to energise a growing environmental movement in Western countries. Progress was made in reducing local air and water pollution by passing Clean Air and Clean Water Acts, but the risk of global ecological overshoot was downplayed. It was only with the realisation of the dangers of climate change due to CO_2 and other greenhouse gas emissions from human industrial activity, comprehensively highlighted in the First Assessment Report of the Inter-governmental Panel on Climate Change (IPCC) in 1990, that led to the signing of the UN Framework Convention on Climate Change (UNFCCC) at the Rio Earth Summit in 1992. The aim of this treaty was to 'stabilise greenhouse gas concentrations in the atmosphere at a level that would prevent dangerous anthropogenic [human-induced] interference in the climate system'.[9] This led to the Kyoto Protocol, signed in 1997, which only covered some industrialised countries and had little impact, and, eventually, to the Paris Agreement in 2015, in which all countries agreed to limit their emissions so as to limit the increase in global temperature to a maximum of 2°C. However, most governments believed that the emissions reductions necessary could still be achieved through a 'green growth' strategy, to achieve absolute decoupling of emissions from economic growth.

Prosperity without growth

The next publication that led to a widespread questioning of the dominant economic growth model was the book *Prosperity without Growth* by British ecological economist Tim Jackson in 2009.[10] This began life as a report for the UK Sustainable Development Commission, which was an official body (since disbanded) advising the UK government on balancing the needs of society, the economy and the environment. As Jackson likes to emphasise, a report questioning the dominant model of pursuing economic growth as the source of prosperity was so controversial that it had to have a question mark after the title – *Prosperity without Growth?* (The question mark was removed from the title of the subsequent book.) This highlights the close association in mainstream economic and political thinking between economic growth and politically desirable goals of jobs and the ability to purchase desired goods and services.

Jackson and others have highlighted two key counter-arguments to this view. First, that beyond the level of income (say, $15,000 per person) needed to ensure access to the basic requirements of personal security, living space and health care, further increases in income do not bring any further significant increases in people's wellbeing or satisfaction with their lives. He attributes this to what he calls the 'iron cage' of consumerism that drives us to buy more and more things that we don't really need. This is partly driven by the role of consumer goods, such as a nice house or a fast car, in signalling a person's social status. So, as Jackson puts it, we continue to buy goods 'to impress people we don't really like'. Whilst we would actually like to spend more time doing the things that really make us happy, like spending more time with family or friends or engaged in purposeful leisure activities, we spend more and more time working to earn money, so as to keep consuming. (Of course, some people, particularly those with control over how they work, also get fulfilment from their jobs.) The paradox here is that we haven't followed the prescription put forward by British economist John Maynard Keynes in the 1930s that we should take the benefits of rising economic prosperity as shorter working times and more family and leisure time.[11]

The second counter-argument to the mainstream view is that there is strong evidence that we are now reaching or exceeding the limits associated with increasing resource use and levels of pollution including carbon emissions. Though we are still far from a complete understanding, our grasp of how the Earth's key systems operate has improved significantly in the last 40 years. Research by Johan Rockström and colleagues, based at the Stockholm Resilience Centre, has examined earth systems and identified nine 'planetary boundaries' that, if crossed, could generate abrupt or irreversible global environmental changes.[12] These represent current best estimates of 'tipping points', where gradual increases in pressure could lead to dramatic change in systems. They argued that human action has already led to the crossing of planetary boundaries associated with disruption of natural nitrogen cycles (essential for plant growth) and loss of biodiversity, as well as climate change. Strikingly, related research led by Will Steffen at the Australian National University has highlighted the 'Great Acceleration' in a number of trends of human global economic activity, including energy use, water use and fertiliser consumption, as well as population and GDP, since 1950. These trends are matched by trends measuring the impacts on earth systems, including carbon dioxide and methane emissions, as well as ocean acidification and tropical forest losses. The increase in global scale of human pressures and impacts has led these researchers to argue that we have entered a new geological era – the Anthropocene, in which human activity is the primary driver of Earth systems change.[13]

So, can we pull back from these planetary boundaries whilst continuing to enjoy the benefits of economic growth? Jackson has used the IPAT equation to illustrate the scale of the challenge of absolute decoupling of impacts from growth. For small percentage changes, the IPAT equation implies that the annual rate of change of carbon emissions r_I is equal to the sum of the rates of changes of the population r_P, affluence r_A and technology r_T terms:

$$r_I = r_P + r_A + r_T$$

Since 1990, global population has increased at a rate of 1.3 per cent per year, and average per capita income has increased at a rate of 1.4 per cent per year. Historically, technological and industrial changes have led to an annual reduction in carbon emissions per unit income at a rate of −0.7 per cent. This has resulted in a growth rate of carbon emissions of 2 per cent per year. As we saw in Chapter 10, to meet the Paris targets, we need carbon emissions per unit income to fall at a rate of around −7 per cent per year, so that CO_2/\$ in 2050 would be 21 times smaller than the current value.

However, Jackson argues that two factors make achieving the necessary carbon reductions even more difficult. First, if we want to create a fairer world, then incomes would need to grow in developing countries so that they can catch up to those in the richer parts of the world. To meet carbon reduction targets and increase global income so that citizens in developing countries can catch up to average European incomes of 2007, global CO_2/\$ in 2050 would need to be 55 times smaller than the current value. Second, if we assume that average incomes in richer countries keep growing at 2 per cent per year over this period, so that developing countries have to catch up with a moving target, then global CO_2/\$ in 2050 would need to be 130 times smaller than the current value. He argues that the rates of improvement in technology that would be needed to achieve this level of carbon intensity reduction strains the imagination of even technological optimists (unless we rely on unproven 'net negative emissions' technologies, as we saw in Chapter 10).

The case against a continuing focus on economic growth that Jackson and others make is that the reductions in carbon emissions needed to maintain a safe climate are incompatible with continuing economic growth in richer countries, if we want to allow poorer countries to grow their economies to create a more equitable world. In short, one of these objectives – carbon reduction, unlimited economic growth and global equity – has to go, and, as economic growth is no longer deli-vering improvements in wellbeing in richer countries, then that is the objective that should be jettisoned. In the next section, we consider the counter-arguments of the 'green growth' proponents, before we go on to examine what an economy not based on growth might look like.

Green growth

The basic story of economic growth that we have set out in this book is that surges of technological change and associated institutional changes have created surpluses that enabled increases in wellbeing, but that these previous surges were based on increasing use of high carbon energy to provide the services that people need and want. The proponents of 'green growth' argue that a new surge of growth based on low carbon energy sources, energy efficiency improvements and related ideas for less resource use forms of production is not only possible, but is necessary to achieve our carbon reduction targets.

British environmental economists Cameron Hepburn and Alex Bowen looked again at Tim Jackson's calculation of the relations between rates of economic growth and the rates of change of the other terms in the IPAT equation.[14] They argued that what these relations show is that reducing GDP growth to zero would not get us very far in terms of achieving the rates of carbon reduction needed, and would still require carbon emissions per unit income to fall at a rate of –5.6 per cent per year. This would still require a massive change in technological systems to deliver energy services with much lower rates of carbon emissions. Moreover, they argue that, on the basis of the type of evidence that we have looked at, a structural change in technological systems requires high rates of economic growth in order to enable the levels of investment needed to bring about that change. So, rather than doing away with economic growth, Hepburn and Bowen argue that we need to harness economic growth to deliver the changes needed for decarbonisation. They recognise that past surges of growth have been energy- and resource-intensive and accept that the material economy is bounded by the natural ecosystem, but argue that the 'intellectual economy', based on providing more knowledge-intensive and less material-intensive goods and services, can keep growing forever. In this scenario, economies would be based more and more on exchange of 'virtual' goods and services, like electronic books, over the Internet and 'experiences', like concerts, in the real world.

Lord Nicholas Stern, lead author of the influential 2006 Stern Review of the Economics of Climate Change, has gone further to argue that low carbon green growth is the new growth story of the twenty-first century and can deliver a new 'wave of innovation'.[15] Pointing to the scale of the challenge, he argues that an urgent redirection of investment is needed, in particular to build the infrastructure to generate and supply low carbon energy services to cities. This is needed to be done globally within the next ten years, as building the types of cities that we have been doing will lock the world into a high carbon pathway. However, Stern optimistically assumes that this type of low carbon investment would deliver economic benefits in terms of job creation and sustainable rates of economic growth.

Stern is part of the Global Commission on the Economy and Climate which has produced a New Climate Economy Report annually since 2014. These reports set out the scale of the changes needed to the current economic system to put the world on a pathway to low or net zero carbon emissions. These include a gradually increasing carbon price, through carbon taxes or linking together regional carbon trading schemes, approaches to investment in low carbon cities and production processes, and changes to financial systems needed to achieve these.[16] They argue that this will require strong and concerted public policies, but that this action will deliver a range of 'co-benefits', as well as carbon reduction and economic growth. For example, action to rapidly move away from high carbon coal-fired electricity generation will also deliver significant improvements in local air quality, as such a move would also reduce the pollutants that give rise to poor air quality. This is particularly a driver of change in countries like China, which are currently suffering greatly from pollutant-driven smogs that are blighting life and health for people in many rapidly developing cities. Other 'co-benefits' could include increasing energy security, as renewable energy technologies are more local and do not need imports of fuel from unstable parts of the world, as well as enhancement of natural ecosystems, such as by reforestation.

Perhaps the strongest argument in favour of green growth, however, is that it works with the grain of current capitalist economic systems. British environmental economist and former senior adviser to Prime Minister Gordon Brown, Michael Jacobs has argued that we don't have the time needed to build a new economic system.[17] He contends that, as governments are still wedded to the current economic systems that rely on continuing high rates of economic growth, promoting green growth is the only feasible option in the short term. The vision of potential economic and social benefits of a green growth pathway is then what enabled governments in both industrialised and developing countries to sign up to the Paris Agreement, and to at least begin to implement measures towards achieving a low carbon transition. In this view, it is possible to create economic value whilst moving away from carbon emissions, and this will require building coalitions of governments, progressive businesses and civil society groups. Jacobs argues that this is the only politically feasible path to achieve change in the short term.

Degrowth and steady state economy

A new group of academics and civil society activists are arguing, on the contrary, that we need a much more radical break with the current political paradigm focused on economic growth, in the form of explicitly advocating a programme of 'degrowth'[18] or a 'steady state economy'. They link the environmental and climate crisis to a wider failure of capitalist economies to deliver increasing wellbeing to the majority of people in industrialised countries. Advocates of this position, like Greek ecological economist Giorgos Kallis, are sceptical of the potential for decoupling of environmental impacts from economic growth, pointing out that a 2 per cent annual growth rate implies a doubling of the size of the economy in 35 years, implying that the logic of exponential growth is incompatible with an economy based on resources from a finite planet.[19] Instead, they advocate a 'transition beyond capitalism'[20] to an economic system based on human-scale goals of caring, community action and activity within environmental limits. Though they make clear that this programme is not about targeting a particular rate of reduction of the value of GDP within the current system (which would have severe negative consequences), they accept that a reduction on the scale of productive activities, as measured by GDP, would be a likely outcome.

As we discussed in Chapter 3, the idea of a steady state economy was first proposed by US ecologist economist Herman Daly, a former student of Georgescu-Roegen, in the 1970s.[21] He argued that the flow or 'throughput' of energy and materials into and out of the economy should be maintained at a steady state which is within the capacity of the biosphere to provide those resources and assimilate the wastes produced. Unfortunately, even though the throughput of energy and materials has stabilised in some industrialised countries due to efficiency improvements, this is usually at a level above their fair share of the Earth's resources.[22] A steady state economy would clearly look very different to present growth-oriented economies, though its advocates argue that it would still be consistent with a high quality of life, measured in less material terms. In their book *Enough is Enough*, US and Canadian ecological economists Rob Dietz and Dan O'Neill examine how a steady state economy could be implemented. They propose measures to reduce resource use and inequality, create jobs, reform the monetary system and change the way that we measure progress through new indicators for the economic system, environment and human wellbeing.[23]

Though such a programme seems unlikely to attract widespread political support any time soon, the degrowth and steady state economy thinkers deserve credit for being willing to think outside the box to what an economy not based on economic growth would be like. A number of the practical ideas that they promote, like reduction in working time, a basic income for all and reform of the financial system, could well be part of the mix of measures needed to achieve a sustainable transition to a low carbon economy, as we discuss in the final chapter. The widespread disenchantment with the way that the current economic system is no longer delivering for the majority of people, as evidenced by the votes for Brexit and Donald Trump, is crying out for an alternative positive vision, which is not based on distrust and exclusion of others. In the end, in democratic countries, change comes about through advocating and mobilising support for new ideas and political leaders who can implement them.

Implications for development

Perhaps the biggest challenge for the degrowth movement is what would be the implications of a reduction in productive activities in industrialised countries on developing countries. Currently, just under 10 per cent of the global population, around 700 million people, live in extreme poverty (defined by the World Bank as living on the equivalent of less than $1.90 per day).[24] By some respects, this represents remarkable progress. In 1990, over 1.9 billion people lived in extreme poverty (then set at $1.25 per day). The United Nations Millennium Development Goals, agreed by all UN countries in 2000, set a goal of halving this number by 2015, which was achieved. This was partly due to international efforts to promote access to nutrition, health care, education and human rights, as well as investment in infrastructure and cancellation of unpayable debt owed by highly indebted poor countries to international development banks. However, the biggest contribution to extreme poverty reduction came from the rapid economic growth in China and other East Asian countries after 2000. This was built on a highly energy-intensive growth path, largely fuelled by the expansion of the use of coal for electricity generation and industrial production. This was largely for the production of cheap goods for export to meet consumption demands in richer countries. Thus, the predominant economic growth path over the last two decades has been in the opposite direction to a path of reducing carbon emissions.

In 2015, the UN agreed on a new set of Sustainable Development Goals for 2030.[25] This set a target of eliminating extreme poverty in the world by 2030, as well as developing the capabilities of those currently living in poverty, alongside a range of other social, economic and environmental goals. These included further action to end hunger and malnutrition, and ensure health and wellbeing and access to primary education for all, as well as ensuring access to affordable, reliable, sustainable and modern energy for all, and taking urgent action to combat climate change. These are clearly laudable goals, and the adoption of goals and associated targets and indicators enables political leaders to be held to account on the progress that they are making towards meeting these goals. However, simultaneously meeting all these goals will require new development pathways.

To be fair, the Chinese leadership has recognised that it has to pursue a more sustainable pathway. At the World Economic Forum in Davos in January 2017, China's President Xi Jingpin defended globalisation but also emphasised that future economic growth must be based on greater use of clean energy, higher levels of energy efficiency and protection of natural resources.[26] This is partly a calculation of self-interest, in that, as we have already discussed, moving to low carbon forms of energy would have the additional benefit of reducing the chronic levels of local air pollution that many Chinese cities face. China is also vulnerable to many of the risks of unabated climate change, including the potential for flooding from its major rivers, and droughts affecting crop harvests. Furthermore, China is now a world leader in renewable energy development and manufacturing capacity, and is beginning to scale back its investment in coal-fired power stations as it expands its domestic deployment of renewable energy technologies, including solar and wind power. Hopefully, this position is also driven by a recognition that its previous high carbon growth path is not sustainable for itself going forward or replicable by other developing and emerging economies, without risking climate catastrophe.

Unfortunately, Western political leaders have been slower to adapt their core economic model. The emphasis since the 2008 financial crisis has been to try to return to higher pre-crisis levels of economic growth, largely through austerity measures aimed at reducing public debt. This has meant a reduction in the levels of public investment in and support for low carbon energy technologies (except in the UK where this is dominated by high levels of public support to encourage private investment in new nuclear power stations, at the expense of public support for a range of renewable energy technologies). The arguments presented in this book suggest that this is a short-sighted approach. Long-run economic growth is driven by innovation in new technologies and institutions, the creation of new opportunities enabled by new knowledge and new capabilities, and the disruption of old industries. The historical evidence presented here suggests that high levels of investment in innovation and deployment of low carbon energy technologies could drive a new surge of economic growth. However, this would require a fundamental restructuring of current energy systems, including institutions, business strategies and user practices, as well as technological change. This would then need a longer-term change to a new techno-economic paradigm.

Energy systems transformation

The scale of energy systems change needed to meet stringent climate change mitigation targets has not yet been recognised by most policy makers. The general perception is that these targets can be met without fundamental changes to current economic systems. Much of the initial focus of a low carbon transition has been on decarbonising electricity supply. This is understandable, as there are a range of low carbon electricity generation technologies: renewable energy technologies, nuclear power and the use of carbon capture and storage with coal and gas generation, as well as the potential for electrification of heat and transport service demands, e.g. through heat pumps and electric vehicles. However, in some ways, this masks the scale of the challenge, as well as potentially limiting the benefits in terms of an economic stimulus.

As US political economists John Zysman and Mark Huberty point out, 'growth via a low carbon energy systems transformation requires a self-sustaining pattern of innovation and investment in both the energy sector and the broader economy'.[27] As we have done in this book, Zysman and Huberty draw on the work of Carlota Perez to point out the systemic nature of past periods of economic growth driven by disruptive technologies that enabled new forms of economic value creation.[28] In most cases, just introducing low carbon electricity generation technologies in current energy systems without any other changes is unlikely to create new economic value, since 'green' electrons are indistinguishable from 'brown' electrons. The main exception is the case of decentralised electricity generation technologies, such as small-scale solar photovoltaics, which do not rely on national electricity grids, and so can bring cheap and convenient electricity, both in industrialised countries and to communities in developing countries without fully developed grids. More broadly, Zysman and Huberty argue that policy makers need to go beyond just looking for least cost decarbonisation, but instead should actively promote energy systems transformation pathways that could generate green growth by stimulating opportunities for new economic value creation and capture. This would include the development of 'smart grids' to marry the potential of low carbon and ICT technologies, investment in a range of new renewable energy technologies, and support for energy efficiency improvements. In these areas, public support could stimulate and provide opportunities for further private sector innovation.

The main point here is the scale of the challenge implied by an energy systems transformation. If such a transformation is to create economic benefits similar to transitions that have driven previous surges of economic growth, then it will need to be of a similar scale to those past transitions, and will need to involve changes in institutions, business strategies and user practices as well as technologies. These could include redesigning regulatory frameworks that were developed for large-scale fossil-fuel driven energy systems, moving to selling energy services rather than units of electricity or gas[29] and users becoming 'prosumers', i.e. producers of energy through small-scale generation as well as consumers. Furthermore, as this transformation is being driven for a social goal of mitigating climate change, rather than largely for private gain, then this will require policy makers to steer the direction of change, whilst providing strong incentives for private sector business and finance involvement. In the short term, a green growth strategy may be more likely to enable this, though a more fundamental reconfiguration of current economic systems, along the lines proposed by the 'degrowth' advocates, may be needed to achieve long-term systemic change.

Agrowth

A compromise proposal between the 'green growth' and 'degrowth' positions has been put forward by Dutch environmental economist Jeroen van den Bergh, who argues that we should pursue an 'agrowth' strategy.[30] He is sceptical about the potential of a green growth strategy to deliver the necessary reductions in carbon emissions, and concerned that such a strategy is always likely to prioritise 'growth' over 'green'. However, he also thinks that the degrowth strategy risks just being interpreted as a reduction in the value of GDP, which, without wider systemic change, would lead to high unemployment and reduction in tax revenues to spend on support for the development of low carbon technologies. He argues instead for an 'agrowth' strategy that focuses on delivering improvements in human wellbeing and reductions in carbon emissions, whilst being agnostic about the implications for the rate of GDP growth. This would incorporate both options that would lead to green growth and options that would reduce GDP growth, provided that they delivered desirable outcomes, such as higher social equity or better health care. He contends that national governments and international organisations, such as the IMF, OECD and World Bank, should therefore reduce the role that they give to assessing policy outcomes by their impacts on GDP growth.

This is a sensible and non-dogmatic position that avoids having to nail your colours to the mast in terms of being in favour or opposed to a focus on GDP growth. However, there may still be perceived trade-offs between carbon reductions and wellbeing improvements. Policies are usually judged against what would have happened otherwise, if the policy had not been implemented. If a proposal would reduce carbon emissions but also reduce wellbeing in the short term, say by increasing energy prices for consumers, then it may be rejected. The political challenge is that any policy that generates losers as well as winners is likely to be strongly opposed by the losers. This means that an appropriate mix of policies will be needed, as well as, crucially, clear vision for the potential benefits to society of a low carbon transition.

An interesting and appealing vision for a new economics, incorporating the 'agrowth' argument, has recently been put forward by British 'renegade economist' Kate Raworth in her book *Doughnut Economics*.[31] Combining the idea of planetary boundaries that set upper ecological limits on the scale of economic activities with the need to deliver minimum social standards for all, in terms of access to energy, clean water, health and social and political rights, Raworth argues that we need to live within the 'doughnut' of a safe and just operating space for humanity. She proposes that twenty-first-century economists should recognise that the economy is embedded within society and the natural world, acknowledge the social aspects of human nature and embrace complexity and evolutionary thinking (as we have done in this book). She contends that this would provide a better guide to action that could deliver social and environmental goals than a continuing political focus on achieving high rates of economic growth.

So, in the final chapter, we look at what this vision and policies might look like, drawing on insights from the green growth, degrowth and agrowth approaches, as well as the lessons from our study of historical technological paradigm changes.

Notes

1 Stern, N. (2016).
2 Kallis (2017a, 2017b).
3 Van den Bergh (2011, 2017).
4 Malthus (1798).
5 Ehrlich and Holden (1971). Strictly speaking, this is an identity, rather than an equation, as both sides of the equation actually contain the same information, just presented in a different form.
6 Meadows et al. (1972).
7 Freeman (1973).
8 Turner (2008, 2014).
9 UN (1992).
10 Jackson (2009/2017).
11 Keynes (1928/1931).
12 Rockström et al. (2009).
13 Steffen et al. (2015).
14 Hepburn and Bowen (2013).
15 Stern, N. (2012, 2015, 2016).

16 Global Commission on the Economy and Climate (2014).
17 Jacobs (2017).
18 D'Alisa et al. (2015); Kallis (2017b).
19 Kallis (2017b).
20 D'Alisa et al. (2015), p. 11.
21 Daly (1977/1991).
22 O'Neill (2015).
23 Dietz and O'Neill (2013).
24 Cruz et al. (2015).
25 UN (2015b).
26 Stern, N. (2017).
27 Zysman and Huberty (2011), p. 12.
28 Zysman and Huberty (2011, 2013).
29 Hannon et al. (2013, 2015).
30 Van den Bergh (2011, 2017).
31 Raworth (2017).

12

CAN WE RISE TO THE CHALLENGE?

Introduction

This book has argued that to understand the relations between energy and economic growth, and what this implies for future energy and economic systems change, we need to combine insights from two areas of thought: evolutionary economics and ecological economics. The evolutionary approach has taught us how technological and economic systems have evolved to deliver five surges of economic growth over the last 250 years, as new innovations replace older systems that are facing diminishing returns. We have also seen that these surges have been driven or closely linked to the use of new sources of energy and improvements in efficiency of their conversion to provide useful work and desired energy services. This means that we need to take seriously the ecological approach that shows the dependence of economic growth on inputs from ecological systems, as well as on the growth of knowledge and institutional capacity.

However, these two approaches have largely operated outside the mainstream of thinking on how economies change and grow at the macro scale. Moreover, they have largely been developed in separate areas of thinking, with little interaction or exchange of ideas between them. This book hopes to start bridging that gap. Though we may not have done justice to the richness of thinking in these two areas, we hope that we have demonstrated that there are synergies between them that suggest ways of usefully bringing ideas together. In particular, both approaches argue that complex systems change comes about through processes of positive and negative feedbacks that reinforce or dampen drivers of change. The evolutionary approach tells us how technological change interacts with wider changes in institutions and financial practices to create revolutions in economic systems and dominant pathways of progress, and identifies the importance of spillovers from these changes in creating new uses and new opportunities. The ecological approach emphasises the importance of substitution of energy-using machines for labour as one of the positive feedbacks driving economic growth, and the importance of considering net energy returns on investment, as well as other potential planetary boundaries.

We think that bringing these two sets of ideas together can provide insights on the daunting environmental, social and economic challenges that the world is now facing. In particular, global energy systems need to be transformed to avoid the worst environmental and social impacts of climate change, whilst maintaining economic development for poorer and emerging economies and reducing social inequalities within and between countries. The evolutionary approach strongly suggests that this could bring economic benefits if the lock-in to current systems and ways of thinking can be overcome. The ecological approach reminds us that technological change is not a 'get out of jail free' card, and that economic systems need to be redesigned so that they are consistent with ecological constraints and focused on delivering human wellbeing. There are clearly tensions between the world views of these two approaches, but the hope is, that by bringing them together, this would suggest new and creative solutions to these challenges.

So, we will end the book by proposing eight key elements for a new pathway to prosperity that build on the evolutionary and ecological economics insights of the range of thinkers and ideas presented here. These will obviously require further development and questioning, but the scale and urgency of the challenges mean that we need to start implementing these or similar ideas to see what works and what needs further critical examination.

Need for a low carbon energy transformation

The economic implications of transforming national and global energy systems so as to mitigate the risks of human-induced climate change are not the only consequences to be assessed when judging what actions we should take, but they are currently amongst the most politically prominent. A minority, but including those now with political power in the US, still want to deny that human actions can affect significantly the climate of our planet. Are we really that powerful? Well, yes, scientists judge we are already exceeding planetary boundaries for maintaining a 'safe operating space' for human life. Based on the study of past technological and economic surges, we would argue that this is the result of scientific knowledge, aligned with technological and institutional capabilities, that has so greatly enhanced our ability to harness natural resources and disrupt natural systems. Of course, scientific knowledge is not infallible, and proceeds on the basis of the 'primary of doubt'.[1] However, it seems ironic, to say the least, to enjoy the fruits of technological advances based on scientific understanding, but to deny the implications of that understanding when it conflicts with political beliefs based on unlimited progress or the need for a smaller role for government in people's lives. I would argue that we have to proceed on the basis of the best scientific understanding of the impacts of human activity on climate and other earth systems. This concludes, beyond reasonable doubt, that continuing to base our economies on high carbon energy systems risks irreversible changes to the Earth's climate, adversely impacting on food and water supplies, frequency of droughts, heatwaves and extreme weather events, potentially leading to social catastrophes in terms of starvation, health impacts, conflicts and mass migration.[2]

So, if we accept the need for a transformation in our economic systems so as to put us on a pathway to an economy and society based on low or zero carbon energy systems, then it seems reasonable to ask what would be the economic consequences of such a transformation? Can we achieve a transformation whilst maintaining or enhancing what most people would regard as a prosperous society? As we have seen, there remain significant differences in the world views of academics and activists working in this area. Some optimists believe that investment in the development and deployment of low carbon energy technologies, particularly renewables, will drive a new surge of economic growth, similar to the past surges that we have examined in this book. Unfortunately, it is far from clear that these technologies have properties similar to those of other technologies that drove these surges, in terms of generating wider economic spillovers. Nevertheless, investment in low carbon technologies will likely create jobs and will enhance energy security, as well as contributing to reducing carbon emissions, and could deliver a boost to economic growth in the short term. If we take an 'agrowth' perspective, then this is not a bad thing, provided that it is linked to measures to reduce wasteful consumption and ensuring that the surplus generated is used to stimulate further low carbon investment. On the other hand, advocates of a 'degrowth' position would argue that a more radical change in the current economic paradigm is needed, away from a focus on economic growth towards an approach based on 'living well within limits'.[3] This would require reorganising political and business activity to prioritise satisfying human needs and measures of social wellbeing. Given the lack of a convincing case for unbounded economic growth on a finite planet, and the fact the economic growth no longer seems to be delivering increasing wellbeing in industrialised countries, then this could be the only viable long-term strategy for maintaining wellbeing whilst reducing carbon emissions to within the safe operating space for humans on this planet.

Elements of a new pathway to prosperity

This then suggests the basis for a new pathway to prosperity: stimulating a new techno-economic surge based on investment in low carbon and resource efficient technologies whilst, at the same time, accepting the need to break away from the current economic model based on financialisation and high levels of consumption to one based on promoting wellbeing whilst reducing social inequality and environmental impacts. But are these two goals not in contradiction? From an evolutionary perspective, we have to start from where we are now, but we can also support alternatives in niches, where they can grow and develop until they are in a position to challenge the dominant paradigm. Critics may argue that this approach would be either too slow to prevent climate disaster or too radical to gain political acceptance, but I would argue that only an approach that promises short-term individual benefits as well as long-term social benefits is likely to have a chance of being politically feasible and to some degree effective in climate terms.

So, what would be the elements of this new pathway? I would identify eight key elements:

- setting a mission promoting 'green' and 'low carbon' as the direction for a new surge of innovation;
- putting a price on carbon emissions through a carbon tax or trading scheme;
- investing in redesigning cities to be smart, low carbon and liveable;
- reducing wasteful consumption by introducing smart, circular economy ideas;
- redefining work by reducing working hours and introducing a basic income;
- reforming the financial system so that it serves the need of the productive economy;
- enhancing democracy so that people have more say over the decisions that affect their lives;
- developing a new economics based on promoting social and ecological justice.

I should stress that none of these ideas is original, and the reader is invited to follow the links in the references to explore more details of these proposed ideas. Of course, it may not be possible to achieve all of these together, but at least it seems like a good idea to try. As a good academic, I have to point out that all of these elements need further research but, most of all, they need to be tried to see what works and what doesn't. When we accept that the current economic system is not able to deliver both social justice and living within environmental limits, then that opens the space, and necessity, for trying to design an economic system that could deliver these.

Low carbon mission

The first step is to accept that government can have a positive role in society. Here, the work of Mariana Mazzucato is helpful in pointing out the role of public support in helping to support the development of many key technologies, such as the Internet, GPS, voice recognition and touch screens, that we usually think of as being driven solely by the private sector.[4] More broadly, she highlights the work of Karl Polanyi, showing that markets are not independent identities, but are 'embedded' in wider institutional structures of law and culture that need to be maintained by government action.[5] An evolutionary perspective emphasises that the main values of markets lie in their ability to coordinate many independent decisions and to promote innovation through encouraging 'creative destruction'.[6]

Whilst accepting that this places limits on government action to promote particular technological options, it is consistent with government providing a strategic direction for economic evolution, based on promoting particular socially desirable outcomes. A mission to promote innovation and systemic change in the direction of low carbon and energy efficient technologies could thus provide a strategic direction for investment and innovation.

Mazzucato and Carlota Perez, the developer of the idea of techno-economic surges of growth, have outlined what a low carbon innovation mission should look like.[7] They emphasise that this needs to include 'direct mission-oriented investment in research, development and tax subsidies that make it clearly profitable to invest in renewable energies, special materials, conservation, recyclability, productivity of resources, and so on'.[8] They argue that this would create jobs in high-quality, high-value manufacturing and service areas, and, together with other policies to promote more personalised but less intensive forms of consumption, could also form the basis for a new model of development for developing countries. They also support reforms to current taxation and financial systems similar to those that we will discuss below.

Their argument is framed in terms of promoting 'green growth', which should help to make it politically attractive to governments. In line with the arguments that we have put forward that, on their own, low carbon technologies are not likely to drive a new surge of innovation, Mazzucato and Perez emphasise that a green revolution would need to include, alongside renewables, moving to a 'circular economy' based on services, rental and maintenance and recycling; redesigning transport systems and cities; and promoting health and education. The evolutionary perspective suggests that promoting innovation in these directions would create synergies and positive feedbacks. However, as with any fundamental change, there would likely be losers amongst the current dominant industries based on intensive use of resources and 'planned obsolescence' of products to encourage further consumption. Hence, it will also be necessary to create public support for these changes, by emphasising the potential benefits to individuals as well as society.

Carbon pricing

The second element is the more widespread adoption of instruments to price carbon emissions. Economists of all stripes agree that, in a market economy, if a side effect of productive activity, like carbon emissions, is not included in the price of the resulting good or service, then it will be over-exploited. In mainstream economic jargon, this is known as an externality, and there is a strong rationale for 'internalising the externality' by putting a price on the emissions to incentivise the producer or consumer to reduce the emissions. There are broadly two ways to do this, either by putting a tax on emissions, or by creating a trading scheme with incentives to reduce emissions. The European Union tried to introduce a carbon tax in the 1990s, but failed to get agreement amongst its member states to do this, as it was (and is) politically difficult to introduce a new tax. Instead, it introduced a carbon Emissions Trading System (ETS), beginning in 2005. Under this system, an overall cap is introduced on emissions from industrial plants (covering around 45 per cent of total emissions), which receive or have to buy emissions allowances. The total number of allowances available is determined by the overall cap which reduces annually. If a firm would be above its individual cap, then it can buy additional allowances from another firm that would be below its cap. In this way, the scheme aims to incentivise emissions to be reduced where it costs least to do this.

Unfortunately, though a trading scheme works fine in economics textbooks, it is more difficult to successfully implement in practice. The European ETS has so far had relatively little impact on reducing carbon emissions. The annual caps were set too high, particularly when economic activity slowed after the 2008 financial crisis, and so the price of allowances that firms had to buy remained low. In early 2017, the price of allowances was around 5 euros per ton of CO_2. This is nowhere near high enough to incentivise firms to make deep cuts in their carbon emissions.

Beyond their lack of success in practice, critics have likened the process of buying allowances to the medieval practice of sinners buying 'indulgences' to atone for their sins.[9] They argue that bringing the incentive for reducing carbon emissions into the financial system means that it becomes just another commodity to be bought and sold, with the potential for creating new financial instruments based on trading carbon futures. Whilst this supports the need for reform of the financial system, as we discuss below, it does not imply that the idea of carbon pricing should be abandoned. In anything like the current economic system, there will be an incentive to overuse any resource that doesn't have a price on it or is otherwise legally regulated. So, some form of carbon pricing would seem to be essential as part of a wider package of measures.

Moreover, carbon pricing (and resource pricing more generally) has the additional benefit of providing a revenue stream for governments to support incentives for innovation and deployment of low carbon options, as well as for other public services. This is important in a time of relatively high levels of national public debt, though, of course, there will be resistance to any new form of taxation, particularly from those industries that would likely be most affected. Fortunately, only a few energy-intensive industries, such as cement, steel and aluminium production, would be particularly affected by an upstream carbon price, and these could be compensated without adversely affecting the effectiveness of the carbon price as a whole.[10]

Again, China is taking the lead by launching a national carbon emissions trading scheme in 2017, beginning with its coal-fired power plants. It is aiming to learn from the relative lack of success of the European ETS, as well as the lack of agreement in the US for a similar national scheme (though regional trading schemes in the Northeastern states and in California are still going ahead). Of course, in some respects, it is easier to push through such a scheme in a country like China without a critical democratic system, but this raises wider questions of the ability to maintain public support necessary for an energy systems transformation without a functioning democratic system.

Implementing a carbon price could be part of a wider ecological tax reform. This has been promoted for many years by environmentalists as a way of shifting taxation away from 'goods' like labour that society wants to encourage and on to 'bads' like pollution that society wants to restrict.[11] This could promote the creation of jobs, as well reducing environmental impacts, and could be particularly important in a time of high levels of tax avoidance by large multinational firms. Again, though, strong political will and public support would be needed to overcome resistance from those who could lose out from this type of tax reform.

Greening of cities

A key site of growth enabled by the fourth oil-based surge has been the growth of cities. Now, over half of the world's population, nearly 4 billion people, live in cities, up from 750 million people in 1950. People want to live in cities, as that is where most of the economic activity takes place, and hence that is where the majority of carbon emissions come from. Cities are also the site of cultural activity and the density of interactions supports the potential for innovation. By 2050, another 2.5 billion people could be living in cities. As the infrastructure to support this is likely to be long-lived, the infrastructure investment decisions made in the next few years will determine the economic and social development of people living in those cities, as well as the pathway of carbon emissions.[12]

The footprint of cities depends on choices made, and so can vary widely. For example, the US city of Atlanta covers more than 25 times the land area of the Spanish city of Barcelona, which has a similar population. As a result of the higher demand for road transport, as well as the high consumption lifestyles of most Americans, Atlanta's CO_2 emissions per person are ten times higher than those of Barcelona.[13]

This suggests the need for better urban planning, focusing on improving people's quality of life, so that opportunities for living and working are placed closer together, and the needs of pedestrians, cyclists and those taking public transport are given higher priority than car users. This would also help to improve local air quality, improving health outcomes. Local government clearly has a key role to play in enabling this, but more effort needs to be put into overcoming the 'institutional gap' preventing wider democratic input into infrastructure investment decision making.[14]

Reducing wasteful consumption

The green growth literature has tended to focus on the need for innovation and deployment of low carbon energy technologies and infrastructure investment, but the scale of the transformation needed to reach climate change targets has to encompass the whole of the economy. As we have seen, increases in final energy use associated with past surges of economic growth have been driven either by increases in primary energy (exergy) input or by improvements in efficiency of energy conversion. In Chapter 10, we argued that decarbonisation of energy supply needs to be accompanied by improvements in energy productivity, i.e. improving the economic value per unit of energy used. Improving energy productivity can be achieved by enhancing the service provided per unit of energy or by enhancing the value associated with that service delivery. This could include relatively simple energy efficiency improvements, such as insulating buildings or improving fuel economy of vehicles, or more structural changes, such as moving to electric vehicles or smarter, more compact cities.

The energy productivity of industrial systems could be improved by implementing so-called circular economy ideas. This involves designing products for recycling, reuse or remanufacturing, as well as new business models, such as leasing vehicles or tools. These ideas are important as they reduce the amount of new resources that are needed in order to provide the services that people want. Our current economic system is still largely based on a linear throughput model, in which new products are manufactured from virgin resources, which quickly end up as waste after one-time use. This may have been appropriate for past economic surges, in which resources were cheap and labour was relatively expensive, and so there was a focus on improving labour productivity. Now that resources are becoming scarce and environmental impacts associated with throughput are becoming clearer, there is a need to focus on energy and resource productivity. In addition to reducing environmental impact, new circular economy business models can enhance the value delivered to users. Here, circular economy ideas overlap with the so-called 'gig' economy, in which more flexibility of provision is enabled by providers using their own spare time and resources, e.g. Uber drivers or Airbnb hosts. Of course, problems with these types of business models have been pointed out by many, as the owners of the platforms enabling this type of service provision extract a 'rent' from each transaction, without providing the benefits that would normally accrue to an employee, such as paid sickness or maternity leave.

This illustrates a wider problem. Modern economic systems are based on increasing consumer demands by providing new or better services that people want to buy. If the demand for these services is not apparent, then different forms of advertising are available to stimulate this demand, and social pressures create expectations of the next 'must have' thing. This is a form of the rebound effect that we encountered in Chapter 10. If a new business model encourages further consumption, e.g. Airbnb encourages people to take more holidays, then the improvement in energy productivity may be taken back in this increased consumption. This suggests the need to look more directly at reducing excessive consumption demand. Limiting the amount of carbon emissions that a person or business could emit, through a carbon tax or cap and trade scheme, would help to limit the amount of carbon-intensive consumption that they could undertake.

However, there is likely a need to go further than this and look at ways to limit any form of wasteful consumption. For example, in richer countries, roughly a third of food produced is lost due to inefficient supply chains or is wasted by retailers or consumers.[15] Any policy moves to restrict consumption levels, though, challenge dominant ideas of consumer sovereignty as well as the influence of large retailers and producers. This is likely to require more systemic changes to the current economic system, such as moving away from increasing GDP as the predominant political focus. UK ecological economist Tim Jackson has proposed moving to a 'Cinderella' economy, in which more value is placed on currently economically marginalised pursuits of caring professions, craft production and cultural activities, which are less material and resource intensive but contribute significantly to increasing human wellbeing.[16]

New forms of work

A further key challenge here is the need for meaningful work for people. Under the current economic system, paid employment is the main means of people securing resources for themselves and their family. (Land and property owners can also extract 'rent' from the ownership of these assets.) The focus on improving labour productivity has usually been achieved through substituting energy-using machines for manual labour, potentially reducing the number of paid jobs in an industry. However, provided that this leads to economic growth that involves the generation of new value-creation opportunities that create new jobs in the economy, then overall levels of employment can be maintained. Classical economists once thought that economies would naturally create full employment in this way, by finding a level of wages at which workers were willing to provide their labour. Some neo-classical economic models still make this assumption. However, in the 1930s, British economist John Maynard Keynes explained the high levels of unemployment in the Great Depression in the UK and the US as a result of lack of 'aggregate demand' in the economy, i.e. because of economic uncertainty, people were not willing to spend enough to create the demand needed to ensure full employment. He advocated government spending as one means to overcome this lack of aggregate demand. Keynes's ideas were part of the mainstream of economic thinking during the golden age of economic growth in the 1950s and 1960s, but fell out of favour from the 1980s as the more free market economic ideas of Friedrich Hayek and Milton Friedman held sway. Keynes's ideas were rapidly rediscovered after the 2008 financial crisis, as governments bailed out failing banks and instituted economic stimulus packages to promote aggregate demand. The proponents of green growth argued, successfully in some cases, such as South Korea and parts of the US package, that investment in clean technologies and energy efficiency measures should be part of a green economic stimulus.[17] The long-term potential for the creation of new green jobs is still debated, however, as this economic model still relies on the creation of new economic activities to provide these jobs.[18]

A second related challenge is that of increasing automation. Authors such as MIT researchers Erik Brynjolfsson and Andrew McAfee have argued that the process of automation of many economic activities through the 'digitisation of just about everything' will lead to many current or professional as well as manual jobs being replaced without corresponding new jobs being created.[19] However, these authors have not considered the interactions between the rise of automation and the green economy. Carlota Perez has argued that circular economy business models, based on repair, reuse, rental and maintenance, would create 'great quantities of jobs'.[20] The question of whether these new jobs would be enough to replace jobs lost through automation has not been fully answered, though.

Interestingly, each of these authors also advocates a form of 'basic income'. This would be an income paid to every citizen of a country to ensure that they have a minimum standard of living. The advantage of this is that it could replace some means-tested benefits, simplifying the welfare system, and addressing the problem of a lack of work due to automation. Critics of a basic income point to the self-worth to individuals associated with having meaningful work, whilst proponents argue that it would create the space for individuals to undertake caring, community work or entrepreneurial ventures without having to worry about providing a minimum standard of living for themselves and their families. Trials of a basic income are now underway in cities in Finland and the Netherlands, which should provide some insights on this. Breaking the link between work and income would be a significant transformation to current economic systems, which may be necessary to address the twin challenges of climate change and automation.

Reforming the financial system

Providing the levels of investment needed will be a key part of a low carbon energy system transformation. However, the current financial system does not seem to be set up to provide this. In previous techno-economic system changes, we have seen how a transition to productive capital from more speculative financial capital after a crash helped to drive the roll-out of new technologies and drive a surge of economic growth. Perez has argued that this usually required institutional changes, as well as a direction for investment. For example, the Glass–Steagall legislation under the 1933 US Banking Act required the separation of commercial and investment banking to prevent financial speculation in the Great Depression, and is credited with helping to enable productive investment along with the New Deal. However, in 1999, after intensive lobbying by the banks, this was repealed by the Gramm–Leach–Bliley Act. At the time, President Bill Clinton claimed that this 'would enhance the stability of our financial services system' by permitting financial firms 'to diversify their product offerings and thus their sources of revenue'.[21] Critics, such as Joseph Stiglitz, former chief economist of the World Bank, on the other hand, argued that this created larger banks deemed too big to be allowed to fail and contributed to the enhanced willingness of these banks to take risks, which ultimately led to the financial crisis of 2008.[22] Following the crash, efforts to restrain speculative investment cumulated in the Dodd–Frank Act of 2012, which created new regulatory oversight of financial institutions, restricted proprietary trading by large banks with their own money, and sought to eliminate 'too big to fail', though again, critics challenged whether this had gone far enough. Ironically, in 2017, the new US President Donald Trump is threatening to repeal the Dodd–Frank

Act. It is notable that key members of the Trump administration, including the new Treasury Secretary Steven Mnuchin, are former members of the investment bank Goldman Sachs. Goldman Sachs has recovered from a perilous financial situation following the crisis, including being forced to pay $5 billion by the US Department of Justice for serious financial misconduct (although no executives were prosecuted), to once again be a leading economic and political actor.

Another approach to encouraging productive rather than purely financial investment is to tax financial trading, as first proposed by US economist James Tobin in 1972. This idea has been taken up by international and non-governmental organisations, like UNICEF and Oxfam, which generally refer to it as a 'Robin Hood' tax, after the British folklore hero who 'robbed from the rich to feed the poor'. In 2010, the European Commission sought to implement this idea, and argued that a Financial Transactions Tax across the European Union could raise €57 billion per year without impeding productive investment. By 2017, though, only ten European countries were still trying to agree to implement this, with negotiations being overshadowed by the continuing economic crisis in Greece.

In the expansion of local renewable electricity generation in Germany, local banks have played a key role by providing long-term finance for renewable energy investment, often owned by those communities. This is seen as aligning with their values of supporting the wellbeing and economic development of these local communities. In turn, these loans were able to be refinanced by the German Development Bank, KfW, which, as we saw in Chapter 7, was set up as part of the Marshall Plan to support the reconstruction of German industry after the Second World War. In contrast, the neoliberal, market-led financial institutions in the UK have been much less supportive of locally owned renewable energy development.[23]

As noted in Chapter 10, the United Nations Environment Programme (UNEP) has argued that a 'quiet revolution' is going on around the world to create new financial and investment practices more aligned to the environmental and social goals of sustainable development. These include the growing use of instruments, such as green bonds, which are raised by public or private lenders to finance environmentally sustainable activities. However, it is likely that a much deeper transformation of the financial system will be needed to support a green transformation.[24] In her testimony to the US Senate on innovation policy, Mazzucato argued that, linked to a mission-oriented programme for supporting green technologies, this would require both public and private lenders to contribute and work together. For example, public development banks, such as the European Investment Bank and the Asian Development Bank, are able to support initial tranches of high risk, which can then persuade private banks to invest in new green technologies.[25] As we argued with colleagues Stephen Hall and Ronan Bolton, financial markets are better understood not as the 'efficient markets' of mainstream economic theory. Instead, they could be seen as 'adaptive markets' consisting of firms and individuals with 'bounded rationality', interacting through evolving systems and networks, that need to overcome behavioural and structural constraints to play their role in financing a transition to low carbon energy systems.[26]

Enhancing democracy

Perhaps the greatest challenge associated with the transformation to a low carbon energy system is the need for popular support, especially, but not only, in countries with democratic governments. There is a view amongst some in the environmental movement that some form of 'benign dictatorship' is necessary in order to force through the political and economic changes to achieve a low carbon transition. I would argue that this is a mistaken view. Even though it is true that countries with authoritarian governments may find it easier to impose costs on certain groups to achieve change 'for the greater good', such as moving whole villages to make way for a hydro-electric dam, they still have to pay attention to what their citizens as a whole deem acceptable. More fundamentally, though, democratic legitimacy is one of the key ways that citizens can hold their government to account. Without this, many individual freedoms and social rights may begin to be eroded. In the end, a sustainable low carbon transition should be about retaining and enhancing those freedoms and rights, so that people are able to articulate what factors are important for their own wellbeing and that of their families and communities.

Moreover, there may be synergies between local energy systems and local democratic control.[27] More local renewable energy and energy efficiency schemes are supported by more local and regional decision-making processes, especially if there are tangible benefits going to local communities. This can also open possibilities for alternative financing, such as crowdfunding, and for new business models for energy supply, such as energy service companies. More local schemes can also enhance the potential for demand management and demand side participation options. However, these local schemes need to be seen as part of an interrelated whole, with large-scale provision and interconnections.

New economics

As we saw in Chapter 10, there are a wide range of low carbon options available, including those for electricity generation and improving energy efficiency that are already or close to being economically viable now. The need for further public and private investment to continue to bring down the costs of these other low carbon technologies through learning, scale, adaptation and network economies is widely accepted. However, limitations on the scale of public support and private investment are usually framed in terms of potential negative impacts on economic growth. This suggests that, in order to be able to gain widespread public and political support for such measures, we need a better understanding of their impacts on economic growth. Unfortunately, as we have argued, mainstream economic thinking is lacking some of the tools needed for this. Furthermore, by hanging on to ideas of equilibrium, market efficiency and perfect competition that are not borne out by understanding of the historical processes of technological and institutional change that have actually driven past surges of economic growth, mainstream economics appears to be part of the problem.[28]

Our final recommendation is therefore the urgent need for new economic thinking to combine the types of qualitative and narrative understandings of the drivers of economic growth that we have presented in this book with more formal quantitative and modelling approaches. I would argue that the types of complexity and evolutionary thinking that underlie the understandings that we have presented here can form the basis for this. Of course, we should not seek to throw out all mainstream economics – just to establish the limits of the validity of approaches based on short-term marginal changes for understanding implications of long-term, systemic changes in economies. We need the deep but partial insights from leading political economists of past generations, including Adam Smith, Karl Marx, Joseph Schumpeter, John Maynard Keynes and Hyman Minsky, to be combined with the insights from a range of more recent ecological, evolutionary and institutional economic thinkers, including Nicholas Georgescu-Roegen, Richard Nelson, Sidney Winter and Elinor Ostrom, to generate this new economics.

As we have seen, it may be possible to combine an understanding of the economy as embedded in ecological systems, with a view of economies as complex adaptive systems.[29] This challenges the mainstream concept of economic value. In mainstream economics, this is simply defined as 'exchange value' – the value of a good or service is what someone is willing to pay for it in a competitive market. Classical economists such as Adam Smith and Karl Marx argued for also considering 'use value', i.e. the value of a good or service should be related to how useful it is. Smith noted that whereas clean water has a high use value, as it is essential for life, it generally has a low exchange value as it is plentiful and easy to obtain. On the other hand, a diamond has little intrinsic use value, except for perception of beauty or signalling the owner's wealth, but a high exchange value, as it is scarce and people are willing to pay highly for it. Mainstream economics denies any value other than exchange value, which is revealed by what people are willing to pay for something. This may be fine for everyday commodities, but breaks down for goods, like clean water or a safe climate, known as 'commons' which belong to societies rather than individuals. It also means that value is limited to what people are able to pay for, so potentially entrenching existing inequalities. If we want to transform our economies so that people can live well within planetary limits, then economics will need to again relate value to usefulness to societies as well as to individuals.

Finally, a new economics will need to revisit the concept of economic growth itself. In this book, we have argued that economic growth is closely related to the use of energy and its efficient conversion from primary inputs to providing useful work. Past surges of economic growth, which have had new energy-using technologies as key inputs, have delivered improvements in human wellbeing. The deep inequalities, both within and between countries, mean that we will continue to need economic improvement for the poorest. Investment in low carbon energy and related resource efficient technologies could help to drive a new surge of 'green growth', if we can design appropriate institutions and incentive structures to support this. However, as we argued in Chapter 11, this needs to go together with a more fundamental redesign of our economic systems. This requires moving away from the economic model currently being followed in rich countries that implies that spending and consuming ever more stuff will make us happy. Thanks to new technologies and institutions designed to support their innovation and deployment, we have greater material wealth than our ancestors could have dreamed of. What we need now is to develop new economic ideas that would enable us to design a system so that that wealth is more evenly shared and innovation is directed towards producing new products and processes that would enhance human wellbeing whilst dramatically reducing our ecological and climate footprint.

It may seem strange to end a book on energy and economic growth with a call for new economics. This is motivated by the view that, in the long run, ideas can challenge the dominance of vested interests, if they motivate action for political change to promote the common good. In the words of Keynes,

> The ideas of economists and political philosophers, both when they are right and when they are wrong, are more powerful than is commonly understood. Practical men, who believe themselves to be quite exempt from any intellectual influence, are usually the slaves of some defunct economist. Madmen in authority, who hear voices in the air, are distilling their frenzy from some academic scribbler of a few years back.[30]

Given the views of some of those currently in authority in parts of the world, at least this academic scribbler hopes that saner voices, drawing on scientific understanding, historical evidence and values of inclusivity and spreading prosperity to the many not the few, will prevail.

Notes

1 Gleick (1992).
2 IPCC (2014).
3 'Living Well Within Limits' is a Leverhulme Research Leadership Award (2017–2022) led by Dr Julia Steinberger at the University of Leeds, see http://lili.leeds.ac.uk/
4 Mazzucato (2013).
5 Polanyi (1944/2001).
6 Schumpeter (1942).
7 Mazzucato and Perez (2014).
8 Mazzucato and Perez (2014), p. 23. See also Mazzucato and Penna (2015).
9 Goodin (1994); Spash (2010).
10 Carbon Trust (2010).
11 Von Weizsäcker and Jesinghaus (1992).
12 Global Commission on the Economy and Climate (2014).
13 Global Commission on the Economy and Climate (2014).
14 Coelho et al. (2014); Hiteva et al. (2017).
15 Priestly (2016).
16 Jackson (2012)
17 Bowen et al. (2009).
18 Zysman and Huberty (2011, 2013).
19 Brynjolfsson and McAfee (2014).

20 Perez (2016).
21 Clinton (1999).
22 Stiglitz (2010).
23 Hall, S. et al. (2016).
24 Naidoo (2016).
25 Mazzucato et al. (2015).
26 Hall, S. et al. (2017).
27 Realising Transition Pathways Engine Room (2015).
28 Jacobs and Mazzucato (2016).
29 Beinhocker (2006); Foxon et al. (2013).
30 Keynes (1936), chapter 24 'Concluding Notes', pp. 383–384.

REFERENCES

Abramovitz, M. (1993) 'Search for sources of economic growth: Areas of ignorance, old and new'. *Journal of Economic History* 53(2), 217–243.

Acemoglu, D. and Robinson, J. (2012) *Why Nations Fail: The Origins of Power, Prosperity and Poverty*. New York: Crown Business.

Aghion, P. and Howitt, P.W. (1998) *Endogenous Growth Theory*. Cambridge, MA: MIT Press.

Allen, R. (2009) *The British Industrial Revolution in Global Perspective*. Cambridge: Cambridge University Press.

Alstone, P., Gershenson, D. and Kammen, D.I. (2015) 'Decentralized energy systems for clean energy access'. *Nature Climate Change* 5, 305–314.

Anderson, K. and Bows, A. (2012) 'A new paradigm for climate change'. *Nature Climate Change* 2, 639–640.

Arapostathis, S., Carlsson-Hyslop, A., Pearson, P.J.G, Gradillas, M., Laczay, S., Thornton, J. and Wallis, S. (2013) 'Governing transitions: Cases and insights from the history of the UK gas industry'. *Energy Policy* 52, 25–44.

Arrhenius, S. (1896) 'On the influence of carbonic acid in the air upon the temperature of the ground'. *London, Edinburgh and Dublin Philosophical Magazine and Journal of Science* (fifth series) April, Vol. 41, 237–275.

Arrhenius, S. (1908) *Worlds in the Making: The Evolution of the Universe*. New York: Harper and Brothers.

Arthur, W.B. (1989) 'Competing technologies, increasing returns and lock-in by historical events'. *Economic Journal* 99, 116–131.

Ayres, R.U. (1997) 'Comments on Georgescu-Roegen'. *Ecological Economics* 22, 285–287.

Ayres, R.U. and Kneese, A.V. (1969) 'Production, consumption, and externalities'. *American Economic Review* 59(3), 282–297.

Ayres, R.U. and Warr, B. (2005) 'Accounting for growth: The role of physical work'. *Structural Change and Economic Dynamics* 16(2), 181–209.

Ayres, R.U. and Warr, B. (2009) *The Economic Growth Engine: How Energy and Work Drive Material Prosperity*. Cheltenham: Edward Elgar.

Ayres, R.U., van den Bergh, J.C.J.M., Lindenberger, D. and Warr, B. (2013) 'The under-estimated contribution of energy to economic growth'. *Structural Change and Economic Dynamics* 27, 79–88.

Beinhocker, E. (2006) *The Origin of Wealth: Evolution, Complexity and the Radical Remaking of Economics*. London: Random House.

Bekar, C.T., Carlaw, K.I. and Lipsey, R.G. (2016) 'General purpose technologies in theory, applications and controversy: A review'. Simon Fraser University Department of Economics Working Paper 16–15.

Bernanke, B.S. (2004) 'The great moderation'. Remarks by Governor Ben S. Bernanke at the meetings of the Eastern Economic Association, Washington, DC, February 20, 2004, www. federalreserve.gov/BOARDDOCS/SPEECHES/2004/20040220/default.htm

Berners-Lee, M. and Clark, D. (2013) *The Burning Question*. London: Profile Books.

Boden, T.A., Marland, G. and Andres, R.J. (2017) *Global, Regional, and National Fossil-Fuel CO$_2$ Emissions*. Carbon Dioxide Information Analysis Center, Oak Ridge National Laboratory, US Department of Energy, Oak Ridge, TN, USA. Available at http://cdiac. ornl.gov/trends/emis/glo_2014.html

Boston Consulting Group (2016) *Global Asset Management 2016: Doubling Down on Data*. Boston, MA: Boston Consulting Group. Available at www.bcgperspectives.com/content/ articles/financial-institutions-global-asset-management-2016-doubling-down-on-data/? chapter=2

Bosworth, B., Burtless, B. and Zhang, K. (2016) *Later Retirement, Inequality in Old Age, and the Growing Gap in Longevity between Rich and Poor*. Economic Studies at Brookings. Washington, DC: Brookings Institution.

Boulding, K.E. (1966) 'The economics of the coming Spaceship Earth'. In Jarrett, H. (ed.), *Environmental Quality in a Growing Economy: Essays from the Sixth RFF Forum*. Baltimore, MD: Johns Hopkins University Press, pp. 3–14. Available at www.ub.edu/prom etheus21/articulos/obsprometheus/BOULDING.pdf

Bowen, A., Fankhauser, S., Stern, N. and Zenghelis, D. (2009) 'An outline of the case for a "green" stimulus'. London: Grantham Research Institute on Climate Change and the Environment/Centre for Climate Change Economics and Policy, Policy Brief, February. Available at http://eprints.lse.ac.uk/24345/1/An_outline_of_the_case_for_a_green_stimulus.pdf

Boyden, S.V. (1992) *Biohistory: The Interplay between Human Society and the Biosphere – Past and Present*. Paris: UNESCO–Parthenon.

BP Statistical Review of World Energy (no date) Oil prices. Available at www.bp.com/en/ global/corporate/energy-economics/statistical-review-of-world-energy/oil/oil-prices.html

Brockway, P.E., Barrett, J.R., Foxon, T.J. and Steinberger, J.K. (2014) 'Divergence of trends in US and UK aggregate exergy efficiencies 1960–2010'. *Environmental Science and Technology* 48, 9874–9881.

Brockway, P.E., Steinberger, J.K., Barrett, J. and Foxon, T.J. (2015) 'Understanding China's past and future energy demand: An exergy efficiency and decomposition analysis'. *Applied Energy* 155, 892–903.

Brown, G. (1994) Speech at an economic seminar, 27 September.

Brynjolfsson, E. and McAfee, A. (2014) *The Second Machine Age: Work, Progress, and Prosperity in a Time of Brilliant Technologies*. New York: W.W. Norton.

Bureau of Economic Analysis (2014) *Measuring the Economy: A Primer on GDP and the National Income and Products Accounts*. Washington, DC: Bureau of Economic Analysis, US Department of Commerce. Available at https://bea.gov/NATIONAL/PDF/NIPA_PRIMER.PDF

Canadell, R., Le Quéré, C., Peters, G. and Jackson, R. (2016) 'Fossil fuel emissions have stalled: Global Carbon Budget 2016'. *The Conversation*, 14 November. Available at https:// theconversation.com/fossil-fuel-emissions-have-stalled-global-carbon-budget-2016-68568

Carbajarles-Dale, M., Raugei, M., Fthenakis, V. and Barnhart, C. (2016) 'Energy Return on Investment (EROI) of solar PV: An attempt at reconciliation'. *Proceedings of the IEEE* 103(7), 995–999.

Carbon Tracker Initiative (2011) *Unburnable Carbon: Are the World's Financial Markets Carrying a Carbon Bubble?* London: Carbon Tracker. Available at www.carbontracker.org/report/carbon-bubble/

Carbon Trust (2010) *Tackling Carbon Leakage: Sector-specific Solutions for a World of Unequal Carbon Prices.* London: Carbon Trust. Available at https://www.carbontrust.com/resour ces/reports/advice/tackling-carbon-leakage-sector-specific-solutions/

Carnahan, W., Ford, K.W., Prosperetti, A., Rochlin, G.I., Rosenfeld, A., Ross, M. et al. (1975) 'Second law efficiency: The role of the second law of thermodynamics in assessing the efficiency of energy use'. *American Institute of Physics Conference Proceedings* 25(1), 25–51, http://doi.org/10.1063/1.30306

Chandler, A.D. (1990) *Economies of Scale and Scope: The Dynamics of Industrial Capitalism.* Cambridge, MA: Harvard University Press.

Chitnis, M., Sorrell, S., Druckman, A., Firth, S.K. and Jackson, T. (2013) 'Turning lights into flights: Estimating direct and indirect rebound effects for UK households'. *Energy Policy* 55, 234–250.

Citi GPS (2013) *Energy Darwinism: The Evolution of the Energy Industry.* London: Citi GPS: Global Perspectives and Solutions. Available at https://www.citivelocity.com/citigps/Rep ortSeries.action?recordId=21

Clark, D. (2012) 'Has the Kyoto Protocol made any difference to carbon emissions?' *The Guardian*, 26 November. Available at www.theguardian.com/environment/blog/2012/nov/26/kyoto-protocol-carbon-emissions

Clark, G. (2014) 'The Industrial Revolution'. In Aghion, P. and Durlauf, S.N. (eds), *Handbook of Economic Growth.* Amsterdam: Elsevier North Holland, Vol. 2, pp. 217–262.

Cleveland, C.J., Constanza, R., Hall, C.A.S. and Kaufmann, R. (1984), 'Energy and the US economy: A biophysical perspective'. *Science* 225, 890–897.

Clinton, W.J. (1999) 'Statement on signing the Gramm–Leach–Bliley Act', 12 November. Available at www.presidency.ucsb.edu/ws/?pid=56922

Coelho, M., Ratnoo, V. and Dellepiane, S. (2014) *The Political Economy of Infrastructure in the UK.* London: Institute for Government. Available at www.instituteforgovernment.org. uk/publications/political-economy-infrastructure-uk

Costanza, R., Low, B.S., Ostrom, E. and Wilson, J. (2000) *Institutions, Ecosystems and Sustainability.* London: CRC Press.

Crafts, N. (2010) 'Explaining the first Industrial Revolution: Two views'. *European Review of Economic History* 15, 153–168.

Cruz, N., Foster, J., Quillin, B. and Schellekens, P. (2015) *Ending Extreme Poverty and Sharing Prosperity: Progress and Policies.* Policy Research Note 3, Development Economics, World Bank. Available at www.worldbank.org/en/research/brief/policy-research-note-03-ending-extreme-poverty-and-sharing-prosperity-progress-and-policies

Csereklyei, Z., Rubio-Varas, M. de M. and Stern, D.I. (2016) 'Energy and economic growth: The stylised facts'. *Energy Journal* 37(2), 223–255.

D'Alisa, G., Demaria, F. and Kallis, G. (2015) *Degrowth: A Vocabulary for a New Era.* London: Routledge.

Daly, H.E. (1977/1991) *Steady State Economics* (2nd Edition). Washington, DC: Island Press.

Daly, H.E. (2015) 'Economics for a full world'. *Great Transition Initiative* (June). Available at www.greattransition.org/publication/economics-for-a-full-world

David, P.A. (1985) 'Clio and the economics of QWERTY'. *American Economic Review* 75(2), 332–337.

David, P.A. (1990) 'The dynamo and the computer: An historical perspective on the modern productivity paradox'. *American Economic Review* 80(2), 355–361.

Denison, E.F. (1979) 'Explanations of declining productivity growth'. *Survey of Current Business* 59(8), Part II, 1–24.

Diamond, J. (2005) *Collapse: How Societies Choose to Fail or Survive.* New York: Penguin.

Dietz, R. and O'Neill, D.W. (2013) *Enough Is Enough: Building a Sustainable Economy in a World of Finite Resources.* San Francisco, CA: Berrett-Koehler Publishers.

Ehrlich, P.R. and Holden, J.P. (1971) 'Impact of population growth'. *Science* 171, 1212–1217.

Eisenhower, D.D. (1961) 'Farewell address to the Nation'. Speech, 17 January.

Ellen Macarthur Foundation (no date) *Circular Economy Overview,* https://www.ellenmaca rthurfoundation.org/circular-economy/overview/concept

Felipe, J. and Fisher, F.M. (2003) 'Aggregation in production functions: What applied economists should know'. *Metroeconomica* 54(2/3), 208–262.

Ferrie, J.E. (2004) *Work, Stress and Health: The Whitehall II Study.* London: Council of Civil Service Unions/Cabinet Office.

Fischer-Kowalski, M. (1998) 'Society's metabolism: The intellectual history of material flow analysis, Part I, 1860–1970'. *Journal of Industrial Ecology* 2(1), 61–78.

Fischer-Kowalski, M. and Hüttler, W. (1998) 'Society's metabolism: The intellectual history of material flow analysis, Part II, 1970–1998'. *Journal of Industrial Ecology* 2(4), 107–136.

Fisher, I. (1933) 'The debt–deflation theory of Great Depressions'. *Econometrica: Journal of the Econometric Society* 1(4), 337–357.

Fouquet, R. (2008) *Heat, Power and Light: Revolutions in Energy Services.* Cheltenham: Edward Elgar.

Fouquet, R. (2010) 'The slow search for solutions: Lessons from historical energy transitions by sector and service'. *Energy Policy* 38, 6586–6596.

Fouquet, R. (2016a) 'Historical energy transitions: Speed, prices and systems transformation'. *Energy Research and Social Science* 22, 7–12.

Fouquet, R. (2016b) 'Lessons from energy history for climate policy: Technological change, demand and economic development'. *Energy Research and Social Science* 22, 79–93.

Fouquet, R. and Pearson, P.J.G. (2006) 'Long run trends in energy services: The price and use of lighting in the United Kingdom, 1300–2000'. *Energy Journal* 25(1), 139–177.

Foxon, T.J. (2010) 'Stimulating investment in energy materials and technologies to combat climate change'. *Philosophical Transactions of the Royal Society A* 368, 3469–3483.

Foxon, T.J. (2011) 'A coevolutionary framework for analysing transition pathways to a sustainable low carbon economy'. *Ecological Economics* 70, 2258–2267.

Foxon, T.J. (2013) 'Transition pathways for a UK low carbon electricity future'. *Energy Policy* 52, 10–24.

Foxon, T.J., Köhler, J., Michie, J. and Oughton, C. (2013) 'Towards a new complexity economics for sustainability'. *Cambridge Journal of Economics* 37, 187–208.

Foxon, T.J., Bale, C.S.E., Busch, J., Bush, R., Hall, S. and Roelich, K. (2015) 'Low carbon infrastructure investment: Extending business models for sustainability'. *Infrastructure Complexity* 2, 4.

FRED Blog (2016) 'The puzzle of real median household income'. Posted on 1 December. https://fredblog.stlouisfed.org/2016/12/the-puzzle-of-real-median-household-income/? utm_source=series_page&utm_medium=related_content&utm_term=related_resources& utm_campaign=fredblog

Freeman, C. (1973) 'Malthus with a computer'. *Futures* 5(1), 5–13.

Freeman, C. (2008) *Systems of Innovation: Selected Essays in Evolutionary Economics.* Cheltenham: Edward Elgar.

Freeman, C. and Perez, C. (1988) 'Structural crises of adjustment: Business cycles and investment behaviour'. In Dosi, G. et al. (eds), *Technical Change and Economic Theory.* London: Pinter [reprinted in Freeman (2008)].

Freeman, C. and Soete, L. (1997) *The Economics of Industrial Innovation* (3rd Edition). London: Routledge.

Freeman, C. and Louçã, F. (2001) *As Time Goes By: From the Industrial Revolutions to the Information Revolution*. Oxford: Oxford University Press.

Geels, F.W. (2002) 'Technological transitions as evolutionary reconfiguration processes: A multi-level perspective and a case study'. *Research Policy* 31(8–9), 1257–1274.

Georgescu-Roegen, N. (1971) *The Entropy Law and the Economic Process*. Cambridge, MA: Harvard University Press.

Gillespie, E. (2009) 'Tesco's "flights for lights" promotion – every little hurts'. *The Guardian*, 6 April. Available at www.theguardian.com/environment/ethicallivingblog/2009/apr/06/tesco-advert-energy-saving-bulbs-flights-greenwash

Gleick, J. (1992) *Genius: The Life and Science of Richard Feynman*. New York: Pantheon Books.

Global Commission on the Economy and Climate (2014) *Better Growth, Better Climate: The New Climate Economy Report*. Available at http://2014.newclimateeconomy.report

Goldemberg, J. (1998) 'Leapfrog energy technologies'. *Energy Policy* 26(10), 729–741.

Goodin, R.E. (1994) 'Selling environmental indulgences'. *Kyklos* 47(4), 573–596.

Grubb, M. with Hourcade, J-C. and Neuhoff, K. (2014) *Planetary Economics: Energy, Climate Change and the Three Domains of Sustainable Development*. London: Routledge.

Grubler, A. (2012) 'Energy transitions research: Insights and cautionary tales'. *Energy Policy* 50, 8–16.

Grubler, A., Wilson, C. and Nemet, G. (2016) 'Apples, oranges, and consistent comparisons of the temporal dynamics of energy transitions'. *Energy Research & Social Science* 22, 18–25.

Haberl, H., Fischer-Kowalski, M., Krausmann, F., Martinez-Alier, J. and Winiwarter, V. (2011) 'A sociometabolic transition towards sustainability? Challenges for another Great Transformation'. *Sustainable Development* 19, 1–14.

Haberl, H., Fischer-Kowalski, M., Krausmann, F. and Winiwarter, V. (2016) *Social Ecology: Society–Nature Relations across Time and Space*. Berlin: Springer.

Hall, C.A.S. and Klitgaard, K. (2012) *Energy and the Wealth of Nations: Understanding the Biophysical Economy*. New York: Springer.

Hall, C.A.S., Powers, R.C. and Schoenberg, W. (2008) 'Peak oil, EROI, investments and the economy in an uncertain future'. In Pimentel, D. (ed.), *Biofuels, Solar and Wind as Renewable Energy Systems: Benefits and Risks*. New York: Springer.

Hall, C.A.S., Balogh, S. and Murphy, D.J. R. (2009) 'What is the minimum EROI that a sustainable society must have?'. *Energies* 2, 25–47.

Hall, S., Foxon, T.J. and Bolton, R. (2016) 'Financing the civic energy sector: How financial institutions affect ownership models in Germany and the United Kingdom'. *Energy Research and Social Science* 12, 5–15.

Hall, S., Foxon, T.J. and Bolton, R. (2017) 'Investing in low carbon transitions: Energy finance as an adaptive market'. *Climate Policy* 17(3), 280–298.

Hammond, G.P and O'Grady, A. (2014) 'The implications of upstream emissions from the power sector'. *Proceedings of the Institution of Civil Engineers – Energy* 167(1), 9–19.

Hannon, M., Foxon, T.J. and Gale, W.F. (2013) 'The coevolutionary relationship between energy service companies and the UK energy system: Implications for a low-carbon transition'. *Energy Policy* 61, 1031–1045.

Hannon, M., Foxon, T.J. and Gale, W.F. (2015) 'Demand–pull government policies to support the product–service system activity: The case of the energy service companies in the UK'. *Journal of Cleaner Production* 108, Part A, 900–915.

Hardin, G. (1968) 'The tragedy of the commons'. *Science* 162(3859), 1243–1248.

Hepburn, C. and Bowen, A. (2013) 'Prosperity with growth: Economic growth, climate change and environmental limits'. In Fouquet, R. (ed.), *Handbook of Energy and Climate Change*. Cheltenham: Edward Elgar.

Hiteva, R., Foxon, T.J. and Lovell, K. (2017) 'The political economy of low carbon infrastructure'. In Kuzemko, C., Goldthau, A. and Keating, M. (eds), *Handbook of International Political Economy of Energy and Natural Resources*. Cheltenham: Edward Elgar.

Hubbert, M.K. (1956) 'Nuclear energy and the fossil fuels'. Presented before Spring meeting of the Southern Division, American Petroleum Institute, San Antonio, TX, 7–9 March. Available at www2.energybulletin.net/node/13630

Hughes, T.P. (1983) *Networks of Power: Electrification in Western Society, 1880–1930*. Baltimore, MD: Johns Hopkins University Press.

IEA (2014) *World Energy Investment Outlook*. Special Report, Paris: International Energy Agency. Available at www.iea.org/publications/freepublications/publication/weo-2014-special-report—investment.html

IEA (2016a) *Energy Technology Perspectives: Towards Sustainable Urban Energy Systems*. Paris: International Energy Agency. Available at www.iea.org/etp/etp2016/

IEA (2016b) *Key World Energy Statistics 2016*. Paris: International Energy Agency. Available at www.iea.org/publications/freepublications/publication/key-world-energy-statistics.html

ILO (2015) *Global Wage Report 2014/15: Wages and Income Inequality*. Geneva: International Labour Organization.

IPCC (2014) *Climate Change 2014: Synthesis Report. Contribution of Working Groups I, II and III to the Fifth Assessment Report of the Intergovernmental Panel on Climate Change* [Core Writing Team, R.K. Pachauri and L.A. Meyer (eds)]. Geneva: IPCC. Available at www.ipcc.ch/report/ ar5/syr/

Jaccard, M. (2005) *Sustainable Fossil Fuels: The Unusual Suspect in the Quest for Clean and Enduring Energy*. Cambridge: Cambridge University Press.

Jackson, T. (2012) 'The Cinderella economy: An answer to unsustainable growth'. *The Ecologist*, 27 July. Available at www.theecologist.org/blogs_and_comments/Blogs/other_blogs/1507111/the_cinderella_economy_an_answer_to_unsustainable_growth.html

Jackson, T. (2009/2017) *Prosperity without Growth: Foundations for the Economy of Tomorrow* (2nd Edition). Abingdon: Routledge.

Jacobs, M. (2017) 'In defence of green growth'. Lecture at ESRC seminar, 'Climate Justice and Economic Growth', University of Manchester, 30–31 January.

Jacobs, M. and Mazzucato, M. (2016) *Rethinking Capitalism: Economics and Policy for Sustainable and Inclusive Growth*. Chichester: Wiley Blackwell.

Jacobsson, S. and Lauber, V. (2006) 'The politics and policy of energy system transformation – explaining the German diffusion of renewable energy technology'. *Energy Policy* 34, 256–276.

Jevons, W.S. (1865) *The Coal Question; An Inquiry Concerning the Progress of the Nation, and the Probable Exhaustion of Our Coal Mines*. London: Macmillan & Co.

Kallis, G. (2017a) 'In defence of degrowth'. Lecture at ESRC seminar, 'Climate Justice and Economic Growth', University of Manchester, 30–31 January.

Kallis, G. (2017b) *In Defense of Degrowth: Opinions and Minifestos*. Open Commons e-book. Available at https://indefenseofdegrowth.com/

Kallis, G. and Norgaard, R. (2010) 'Coevolutionary ecological economics'. *Ecological Economics* 69, 690–699.

Kander, A. and Stern, D.I. (2014) 'Economic growth and the transition from traditional to modern energy in Sweden'. *Energy Economics* 46, 56–65.

Kander, A., Malanima, P. and Warde, P. (2014) *Power to the People: Energy in Europe over the Last Five Centuries*. Princeton, NJ: Princeton University Press.

Keynes, J.M. (1919) *The Economic Consequences of the Peace*. London: Macmillan.

Keynes, J.M. (1928/1931) 'Economic possibilities for our grandchildren'. In Keynes, J.M. (1931), *Essays in Persuasion*. London: Macmillan.

Keynes, J.M. (1936) *The General Theory of Employment, Interest and Money*. London: Macmillan.

KfW (no date) *History of KfW*. Available at www.kfw.de/KfW-Group/About-KfW/Identi tät/Geschichte-der-KfW/

King, M. (2003) Speech given by Mervyn King, Governor of the Bank of England, East Midlands Development Agency/Bank of England dinner, Leicester, 14 October. Available at www. bankofengland.co.uk/archive/Documents/historicpubs/speeches/2003/speech204.pdf

Klein, N. (2015) *This Changes Everything: Capitalism vs. The Climate*. New York: Simon & Schuster.

Kümmel, R. (2011) *The Second Law of Economics: Energy, Entropy and the Origins of Wealth*. New York: Springer.

Kümmel, R., Strassl, W., Gossner, A. and Eichhorn, W. (1985) 'Technical progress and energy dependent production functions'. *Journal of Economics (Zeitschrift für Nationalökonomie)* 45, 285–311.

Lipsey, R.G., Carlaw, K.I. and Bekar, C.T. (2005) *Economic Transformations: General Purpose Technologies and Long Term Economic Growth*. Oxford: Oxford University Press.

Lucas, R.M. (1988) 'On the mechanics of economic development'. *Journal of Monetary Economics* 22(1), 3–42.

McGlade, C. and Ekins, P. (2015) 'The geographical distribution of fossil fuels unused when limiting global warming to 2°C'. *Nature* 517, 187–190.

Maddison, A. (2010) 'Statistics on world population, GDP and GDP per capita, 1–2008 AD', Maddison Project database. Available at www.ggdc.net/maddison/maddison-p roject/home.htm

Malthus, T.R. (1798) *An Essay on the Principle of Population*. London: J. Johnson.

Maréchal, K. (2007) 'The economics of climate change and the change of climate in economics'. *Energy Policy* 35, 5181–5194.

Marx, K. (1867–1894) *Capital: A Critique of Political Economy*, Volumes I, II and III. Hamburg: Verlag von Otto Meissner.

Mathews, J.A. (2011) 'Naturalizing capitalism: The next Great Transformation'. *Futures* 43, 868–879.

Mathews, J.A. (2015) *Greening of Capitalism: How Asia Is Driving the Next Great Transformation*. Stanford, CA: Stanford University Press.

Mathews, J.A. and Reinert, E.S. (2014) 'Renewables, manufacturing and green growth: Energy strategies based on capturing increasing returns'. *Futures* 61, 13–22.

Mazzucato, M. (2013) *The Entrepreneurial State: Debunking Public vs. Private Sector Myths*. London: Anthem Press.

Mazzucato, M. and Perez, C. (2014) 'Innovation as growth policy: The challenge for Europe'. SPRU Working Paper Series SWPS 2014–13, July, University of Sussex. Available at www.sussex.ac.uk/webteam/gateway/file.php?name=2014-13-swps-mazzuca to-perez.pdf&site=25

Mazzucato, M. and Penna, C. (2015) *Mission-oriented Finance for Innovation: New Ideas for Investment-led Growth*. Policy Network. London: Rowman and Littlefield International.

Mazzucato, M., Semeniuk, G. and Watson, J. (2015) 'What will it take to get a green revolution? No more nudging. Pushing on supply and pulling on demand'. Sussex Energy Group, University of Sussex. Available at www.sussex.ac.uk/webteam/gateway/ file.php?name=what-will-it-take-to-get-us-a-green-revolution.pdf&site=264

Meadows, D.H, Meadows, D.L., Randers, J. and Behrens III, W.W. (1972) *The Limits to Growth: A Report for the Club of Rome's Project on the Predicament of Mankind*. New York: Universe Books.

Mokyr, J. (2009) *The Enlightened Economy*. London: Penguin Books.

Mokyr, J. (2016) *A Culture of Growth: The Origins of the Modern Economy*. Princeton, NJ: Princeton University Press.

Morozov, E. (2014) *To Save Everything, Click Here: Technology, Solutionism and the Urge to Fix Problems that Don't Exist*. London: Penguin.

Morris, C. and Pehnt, M. (2012/2016) *Energy Transition: The German Energiewende*. Berlin: Heinrich Böll Foundation, e-book https://book.energytransition.org/

Murmann, J.P. (2003) *Knowledge and Competitive Advantage: The Coevolution of Firms, Technology and National Institutions*. Cambridge: Cambridge University Press.

Murphy, D.J. and Hall, C.A.S. (2011) 'Energy return on investment, peak oil and the end of economic growth'. *Annals of the New York Academy of Sciences* Issue: Ecological Economics Reviews 1219(1): 52–72.

Naidoo, C. (2016) 'The coevolution of financial ecosystems and green transformations'. SPRU, University of Sussex, unpublished.

Nelson, R.R. (2005) *Technology, Institutions and Economic Growth*. Cambridge, MA: Harvard University Press.

Nelson, R.R. (2008) 'What enables rapid economic progress? What are the needed institutions?' *Research Policy* 37, 1–11.

Nelson, R.R. and Winter, S.G. (1982) *An Evolutionary Theory of Economic Change*. Cambridge, MA: Harvard University Press.

Norgaard, R. (1994) *Development Betrayed: The End of Progress and a Coevolutionary Revisioning of the Future*. London: Routledge.

North, D.C. (1990) *Institutions, Institutional Change and Economic Performance*. Cambridge: Cambridge University Press.

O'Neill, D.W. (2015) 'The proximity of nations to a socially sustainable steady-state economy'. *Journal of Cleaner Production* 108, Part A, 1213–1231.

Osborne, R. (2013) *Iron, Steam & Money: The Making of the Industrial Revolution*. London: Pimlico.

Ostrom, E. (1990) *Governing the Commons: The Evolution of Institutions for Collective Action*. Cambridge: Cambridge University Press.

Oxfam (2017) *An Economy for the 99%*. Oxfam Briefing Paper, January. Oxford: Oxfam GB.

Pearson, P.J. and Foxon, T.J. (2012) 'A low carbon industrial revolution? Insights and challenges from past technological and economic transformations'. *Energy Policy* 50, 117–127.

Perez, C. (1983) 'Structural change and the assimilation of new technologies in the economic and social systems'. *Futures* 15, 357–375.

Perez, C. (2002) *Technological Revolutions and Financial Capital: The Dynamics and Bubbles of Golden Ages*. Cheltenham: Edward Elgar.

Perez, C. (2013) 'Unleashing a golden age after the financial collapse: Drawing lessons from history'. *Environmental Innovation and Societal Transitions* 6, 9–23.

Perez, C. (2016) 'Capitalism, technology and a green global golden age: The role of history in helping to shape the future'. In Jacobs, M. and Mazzucato, M. (eds), *Rethinking Capitalism: Economics and Policy for Sustainable and Inclusive Growth*. Chichester: Wiley Blackwell.

Piketty, T. (2014) *Capital in the 21st Century*. Cambridge, MA: Harvard University Press.

Piketty, T. and Saez, E. (2003) 'Income inequality in the United States, 1913–1998'. *Quarterly Journal of Economics* 118(1), 1–39.

Pierson, P. (2000) 'Increasing returns, path dependence, and the study of politics'. *American Political Science Review* 94(2), 251–267.

Plumer, B. (2017) 'Scientists make a detailed "roadmap" for meeting the Paris climate goals. It's eye-opening'. *Vox*, 24 March. Available at www.vox.com/energy-and-environment/2017/3/23/15028480/roadmap-paris-climate-goals

Polanyi, K. (1944/2001) *The Great Transformation: The Political and Economic Origins of Our Time*. Boston, MA: Beacon Press.

Priestly, S. (2016) *Food Waste*. House of Commons Briefing Paper Number CBP07552, August. Available at http://researchbriefings.files.parliament.uk/documents/CBP-7552/CBP-7552.pdf

Raworth, K. (2017) *Doughnut Economics: Seven Ways to Think Like a 21st Century Economist*. London: Random House Business.

Realising Transition Pathways Engine Room (2015) *Distributing Power: A Transition to a Civic Energy Future*. Realising Transition Pathways Research Consortium, University of Bath. Available at http://opus.bath.ac.uk/48114/1/FINAL_distributing_power_report_WEB.pdf

Rockström, J., Steffen, W., Noone, K., Persson, Å., Chapin, III, F.S., Lambin, E. et al. (2009) 'Planetary boundaries: Exploring the safe operating space for humanity'. *Ecology and Society* 14(2), 32. Available at www.ecologyandsociety.org/vol14/iss2/art32/

Rockström, J., Gaffney, O., Rogelj, J., Meinshausen, M., Nakicenovic, N. and Schellnhuber, J.H. (2017) 'A roadmap for rapid decarbonization'. *Science* 355(6331), 1269–1271.

Romer, P.M. (1986) 'Increasing returns and long-run growth'. *Journal of Political Economy* 94(5), 1002–1037.

Romer, P.M. (1994) 'The origins of endogenous growth'. *Journal of Economic Perspectives* 8(1), 3–22.

Sagan, C. (1994) Speech at Cornell University, 13 October.

Samuelson, P.A. (1946) *Foundations of Economics*. Cambridge, MA: Harvard University Press.

Samuelson, P.A. (1948) *Economics: An Introductory Analysis*. New York: McGraw-Hill.

Samuelson, P.A. and Nordhaus, W.D. (2009) *Economics* (19th Edition). New York: McGraw-Hill.

Schot, J. and van Lente, D. (2010) 'Technology, industrialization and the contested modernization of the Netherlands'. In Schot, J., Lintsen, H. and Rip, A. (eds), *Technology and the Making of the Netherlands: The Age of Contested Modernization, 1890–1970*. Cambridge, MA: MIT Press, pp. 485–542.

Schot, J. and Kanger, L. (2016) 'Deep transitions: Emergence, acceleration, stabilization and directionality'. *SPRU Working Paper Series* SWPS 2016–15 (September), University of Sussex. Available at www.sussex.ac.uk/webteam/gateway/file.php?name=2016-15-swps-schot-et-al.pdf&site=25

Schreurs, M. (2015) 'The German green energy transition'. Video, Free University Berlin. Available at www.polsoz.fu-berlin.de/en/polwiss/forschung/systeme/ffu/aktuell/15_okt_schreurs-youtube.html

Schrödinger, E. (1944/2012) *What Is Life?* Cambridge: Cambridge University Press.

Schumpeter, J.A. (1911/1934) *The Theory of Economic Development*. Cambridge, MA: Harvard University Press.

Schumpeter, J.A. (1942) *Capitalism, Socialism and Democracy*. New York: Harper & Brothers.

Schurr, S.H. and Netschert, B.C. (1960) *Energy in the American Economy, 1850–1975: An Economic Study of Its History and Prospects*. Baltimore, MD: Johns Hopkins Press.

Schwab, K. (2016) *The Fourth Industrial Revolution*. New York: Crown Business.

Sieferle, R.P. (1982/2001) *The Subterranean Forest: Energy Systems and the Industrial Revolution*. Cambridge: White Horse Press.

Smil, V. (1999) 'Detonator of the population explosion'. *Nature* 400, 415.

Smil, V. (2010) *Energy Transitions: History, Requirements, Prospects*. Santa Barbara, CA: Praeger.

Smith, A. (1776/1999) *The Wealth of Nations, Books I–III*. London: Penguin Books.

Soddy, F. (1926) *Wealth, Virtual Wealth and Debt: The Solution of the Economic Problem*. New York: E.P. Dutton.

Solow, R. (1956) 'A contribution to the theory of economic growth'. *Quarterly Journal of Economics* 70(1), 65–94.

Solow, R. (1957) 'Technical change and the aggregate production function'. *Review of Economics and Statistics* 39(3), 312–320.

Solow, R. (1987) 'We'd better watch out'. *New York Times Book Review*, 12 July, p. 36.

Sovacool, B.K. (2016) 'How long will it take? Conceptualizing the temporal dynamics of energy transitions'. *Energy Research & Social Science* 13, 202–215.

Sovacool, B.K. and Geels, F.W. (2016) 'Further reflections on the temporality of energy transitions: A response to critics'. *Energy Research & Social Science* 22, 232–237.

Spash, C. (2010) 'The Brave New World of carbon trading'. *New Political Economy* 15(2), 169–195.

Steffen, W., Broadgate, W., Deutsch, L., Gaffney, O. and Ludwig, C. (2015) 'The trajectory of the Anthropocene: The Great Acceleration'. *Anthropocene Review* 2(1), 81–98.

Stenzel, T. and Frenzel, A. (2008) 'Regulating technological change – the strategic reactions of utility companies towards subsidy policies in the German, Spanish and UK electricity markets'. *Energy Policy* 36, 2645–2657.

Stern, D.I. (2011) 'The role of energy in economic growth'. *Annals of the New York Academy of Sciences* Issue: Ecological Economics Reviews 1219(1): 26–51.

Stern, D.I. and Kander, A. (2012) 'The role of energy in the industrial revolution and modern economic growth'. *Energy Journal* 33(3), 127–154.

Stern, N. (2007) *The Economics of Climate Change: The Stern Review*. Cambridge: Cambridge University Press.

Stern, N. (2012) 'Climate change and the new industrial revolution'. Lionel Robbins Memorial Lectures, London School of Economics, 21–23 February. Available at www.lse.ac.uk/GranthamInstitute/news/lord-nicholas-stern-delivers-three-lectures-on-climate-change-and-the-new-industrial-revolution/

Stern, N. (2015) *Why Are We Waiting? The Logic, Urgency, and Promise of Tackling Climate Change*. Cambridge, MA: MIT Press.

Stern, N. (2016) 'The criticality of the next 10 years: delivering the global agenda and building infrastructure for the 21st century'. Speech at Stern Review+10: New opportunities for growth and development, Royal Society, London, 28 October. Available at www.lse.ac.uk/GranthamInstitute/event/the-stern-review-10-new-opportunities-for-growth-and-development/

Stern, N. (2017) 'China is shaping up to be a world leader on climate change'. *Financial Times*, 20 January.

Stiglitz, J. (2010) *Freefall: America, Free Markets and the Sinking of the World Economy*. New York: W.W. Norton.

Tainter, J.A. (1988) *The Collapse of Complex Societies*. Cambridge: Cambridge University Press.

Tainter, J.A. (1996) 'Complexity, problem solving, and sustainable societies'. In Costanza, R., Segura, O. and Martinez-Alier, J. (eds), *Getting Down to Earth: Practical Applications of Ecological Economics*. Washington, DC: Island Press.

Tainter, J.A. and Patzek, T.W. (2012) *Drilling Down: The Gulf Oil Debacle and Our Energy Dilemma*. Berlin: Springer.

Turner, G. (2008) 'A comparison of *The Limits to Growth* with 30 years of reality'. *Global Environmental Change* 18, 397–411.

Turner, G. (2014) *Is Global Collapse Imminent? An Updated Comparison of The Limits to Growth with Historical Data*. Melbourne Sustainable Society Institute Research Paper 4, University of Melbourne. Available at http://sustainable.unimelb.edu.au/sites/default/files/docs/MSSI-ResearchPaper-4_Turner_2014.pdf

Turnheim, B. and Geels, F.W. (2012) 'Regime destabilisation as the flipside of energy transitions: Lessons from the history of the British coal industry (1913–1997)'. *Energy Policy* 50, 35–49.

Tylecote, A. (1992) *The Long Wave in the World Economy: The Present Crisis in Historical Perspective.* London: Routledge.

United Nations (1992) *Framework Convention on Climate Change.* Available at https://unfccc.int/resource/docs/convkp/conveng.pdf

United Nations (2015a) *Paris Agreement.* Paris: United Nations Framework Convention on Climate Change. Available at http://unfccc.int/paris_agreement/items/9485.php

United Nations (2015b) *Sustainable Development Goals.* New York: United Nations. Available at https://sustainabledevelopment.un.org/

United Nations Environment Programme (2015) *The Financial System We Need: Aligning the Financial System with Sustainable Development.* UNEP Inquiry Report. Available at http://unepinquiry.org/publication/inquiry-global-report-the-financial-system-we-need/

United Nations Environment Programme (2016) *The Emissions Gap Report 2016: A UNEP Synthesis Report.* Nairobi: UNEP. Available at www.unep.org/emissionsgap/resources

Unruh, G.C. (2000) 'Understanding carbon lock in'. *Energy Policy* 28, 817–830.

Utterback, J.M. and Abernathy, W.J. (1975) 'A dynamic model of product and process innovation'. *Omega: The International Journal of Management Science* 3(6), 639–656.

van den Bergh, J.C.J.M. (2011) 'Environment versus growth – a criticism of "degrowth" and a plea for "a-growth"'. *Ecological Economics* 70, 881–890.

van den Bergh, J.C.J.M. (2017) 'A third option for climate policy within limits to growth'. *Nature Climate Change* 7, 107–112.

Varoufakis, Y. (2015) *The Global Minotaur: America, Europe and the Future of the Global Economy.* London: Zed Books.

Vivid Economics (2017) *Economic Growth in a Low Carbon World: How to Reconcile Growth and Climate through Energy Productivity.* Report prepared for the Energy Transitions Commission. London: Vivid Economics. Available at www.energy-transitions.org/sites/default/files/carbon-world.pdf

Von Tunzelmann, G.N. (1978) *Steam Power and British Industrialization to 1860.* New York: Oxford University Press.

Von Weizsäcker, E. and Jesinghaus, J. (1992) *Ecological Tax Reform: A Policy Proposal for Sustainable Development.* London: Zed Books.

Warr, B. and Ayres, R.U. (2006) 'REXS: A forecasting model for assessing the impact of natural resource consumption and technological change on economic growth'. *Structural Change and Economic Dynamics* 17, 329–378.

Warr, B. and Ayres, R.U. (2012) 'Useful work and information as drivers of growth'. *Ecological Economics* 73, 93–107.

WID (no date) World Wealth and Income Database. Available at http://wid.world/data/

Wilkinson, R. and Pickett, K. (2010) *The Spirit Level: Why Equality Is Better for Everyone* (2nd Edition). London: Penguin Books.

Wrangham, R. (2009) *Catching Fire: How Cooking Made Us Human.* London: Profile Books.

Wrigley, E.A. (2010) *Energy and the English Industrial Revolution.* Cambridge: Cambridge University Press.

Wrigley, E.A. (2013) 'Energy and the English Industrial Revolution'. *Philosophical Transactions of the Royal Society A* 371, 20110568.

Wrigley, E.A. (2016) *The Path to Sustained Growth: England's Transition from an Organic Economy to an Industrial Revolution.* Cambridge: Cambridge University Press.

Young, A.A. (1928) 'Increasing returns and economic progress'. *Economic Journal* 38(152), 527–542.

Zysman, J. and Huberty, M. (2011) 'From religion to reality: Energy systems transformation for sustainable prosperity'. In Huberty, M. et al. (eds), *Green Growth: From Religion to Reality*. Berkeley Roundtable on the International Economy. Available at www.sustainia. me/resources/publications/mm/From-religion-to-reality.pdf

Zysman, J. and Huberty, M. (2013) *Can Green Sustain Growth? From the Religion to the Reality of Sustainable Prosperity*. Stanford, CA: Stanford University Press.

INDEX

Note: page numbers in italic type refer to Figures; those in bold type refer to Tables.